AMERICAN POLITICAL, ECONOMIC, AND SECURITY ISSUES

NUCLEAR SECURITY

ASSESSMENTS OF MONITORING AND VERIFICATION NEEDS AND IAEA PROGRAMS

American Political, Economic, and Security Issues

Additional books in this series can be found on Nova's website under the Series tab.

Additional e-books in this series can be found on Nova's website under the e-book tab.

AMERICAN POLITICAL, ECONOMIC, AND SECURITY ISSUES

NUCLEAR SECURITY

ASSESSMENTS OF MONITORING AND VERIFICATION NEEDS AND IAEA PROGRAMS

CHARLOTTE R. CAREY
EDITOR

nova publishers
New York

For permission to use material from this book please contact us:
Telephone 631-231-7269; Fax 631-231-8175
Web Site: http://www.novapublishers.com

Library of Congress Cataloging-in-Publication Data

ISBN: 978-1-63117-643-2

Published by Nova Science Publishers, Inc. † New York

CONTENTS

PREFACE

The International Atomic Energy Agency (IAEA)plays a crucial role in supporting U.S. nuclear nonproliferation goals through its safeguards and nuclear security programs. The Department of State (State) coordinates the United States' financial and policy relationship with IAEA. IAEA's safeguards program is designed to detect and deter the diversion of nuclear material for non-peaceful purposes, while the agency's nuclear security program assists countries in improving the physical protection of their nuclear material and facilities. IAEA plans to create an international fuel bank to guarantee the supply of fuel for civilian nuclear power programs. This book examines any challenges that IAEA faces in carrying out its safeguards program; any limitations regarding the nuclear security program; and the status of IAEA's planned nuclear fuel bank.

Chapter 1 – The Defense Science Board Task Force on Assessment of Nuclear Treaty Monitoring and Verification Technologies was established to examine a broad range of questions concerning the capability of the Department of Defense (DoD), the Department of Energy (DOE) and the Intelligence Community (IC) to support future monitoring and verification of nuclear nonproliferation and arms control treaties. The Terms of Reference (TOR) for the study, found in Appendix B, state the tasking. Given the breadth of the topics of interest to our sponsoring leadership and the time and resources available, the Task Force determined to focus on those aspects of the TOR that address what it views as the priority issue—namely, monitoring for proliferation. Assessments of strategies for monitoring nuclear activities in both permissive and non-permissive environments, and of our current technical capabilities and future requirements for successfully implementing those strategies, were made.

Chapter 2 – IAEA plays a crucial role in supporting U.S. nuclear nonproliferation goals through its safeguards and nuclear security programs. The Department of State (State) coordinates the United States' financial and policy relationship with IAEA. IAEA's safeguards program is designed to detect and deter the diversion of nuclear material for non-peaceful purposes, while the agency's nuclear security program assists countries in improving the physical protection of their nuclear material and facilities. IAEA plans to create an international fuel bank to guarantee the supply of fuel for civilian nuclear power programs. GAO was asked to examine (1) any challenges that IAEA faces in carrying out its safeguards program, (2) any limitations regarding the nuclear security program, and (3) the status of IAEA's planned nuclear fuel bank.

GAO reviewed relevant documents and interviewed officials from IAEA and U.S. federal agencies, 15 nuclear nonproliferation experts, and representatives from 16 countries.

In: Nuclear Security
Editor: Charlotte R. Carey

ISBN: 978-1-63117-643-2

Chapter 1

ASSESSMENT OF NUCLEAR MONITORING AND VERIFICATION TECHNOLOGIES*

Defense Science Board

EXECUTIVE SUMMARY

The Defense Science Board Task Force on Assessment of Nuclear Treaty Monitoring and Verification Technologies was established to examine a broad range of questions concerning the capability of the Department of Defense (DoD), the Department of Energy (DOE) and the Intelligence Community (IC) to support future monitoring and verification of nuclear nonproliferation and arms control treaties. The Terms of Reference (TOR) for the study, found in Appendix B, state the tasking. Given the breadth of the topics of interest to our sponsoring leadership and the time and resources available, the Task Force determined to focus on those aspects of the TOR that address what it views as the priority issue—namely, monitoring for proliferation. Assessments of strategies for monitoring nuclear activities in both permissive and non-permissive environments, and of our current technical capabilities and future requirements for successfully implementing those strategies, were made.

A Nuclear Future Unlike the Past and the Need for a Monitoring Paradigm Shift

Too many factors have changed, and are changing from our historic basis and experience developed throughout the Cold War. The list of factors that should give national leadership pause for concern is extensive, and includes the following:

- The actual or threatened acquisition of nuclear weapons by more actors—with a range of motivations, capabilities, and approaches—is emerging in numbers not seen since the early days of the Cold War. Many of these actors are hostile to the U.S. and

* This is an edited, reformatted and augmented version of a Task Force Report dated January 2014.

its allies, and of greater worry, they do not appear to be bound by established norms nor are they deterred by traditional means;

- In some cases, nuclear forces are seen as the most affordable and effective alternative to deter superior conventional forces; i.e., nuclear weapons are viewed as a legitimate warfighting capability, especially if vital domestic or regional security interests are threatened;
- Fundamental nuclear knowledge is widespread and know-how increasingly accessible.
    - At the same time, ubiquitous information access and widespread observational tools are increasing inherent transparency;
    - However, recognition of such increased transparency by potential or actual proliferants incentivizes the employment of more sophisticated methods of denial and deception;
- The pathways to proliferation are expanding to include networks of cooperation among nations and actors who would otherwise have little reason to do so;
- The growth in nuclear power worldwide offers more opportunity for "leakage" and/or hiding small programs.

In this unfolding nuclear future, monitoring to support treaties is but one part of the overall monitoring requirement that should be driven by monitoring for proliferation. This broader scope presents challenges for which current solutions are either inadequate, or more often, do not exist. Among these challenges are monitoring of:

- Small inventories of weapons and materials, even as low as a single "significant quantity of fissile material";
- Small nuclear enterprises designed to produce, store, and deploy only a small number of weapons—intended as a proliferant's end goal, or as the first steps to achieve larger inventories or more sophisticated capabilities;
- Undeclared facilities and/or covert operations, such as testing below detection thresholds, or acquisition of materials or weapons through theft or purchase;
- Use of non-traditional technologies, presenting at best ambiguous signatures, to acquire both materials and components;
- Theater nuclear forces and associated doctrine, exercises, and training complicated by the use of mobile, dual-use delivery systems;
- Many more players to whom access by the U.S. or its allies will be limited or extremely difficult, some of whom will be globally networked with global access to relevant science and technology.

In short, for the first time since the early decades of the nuclear era, the nation needs to be equally concerned about both "vertical" proliferation (the increase in capabilities of existing nuclear states) and "horizontal" proliferation (an increase in the number of states and non-state actors possessing or attempting to possess nuclear weapons). These factors, and others discussed more fully in the body of this report, led the Task Force to observe that monitoring for proliferation should be a top national security objective—but one for which the nation is not yet organized or fully equipped to address.

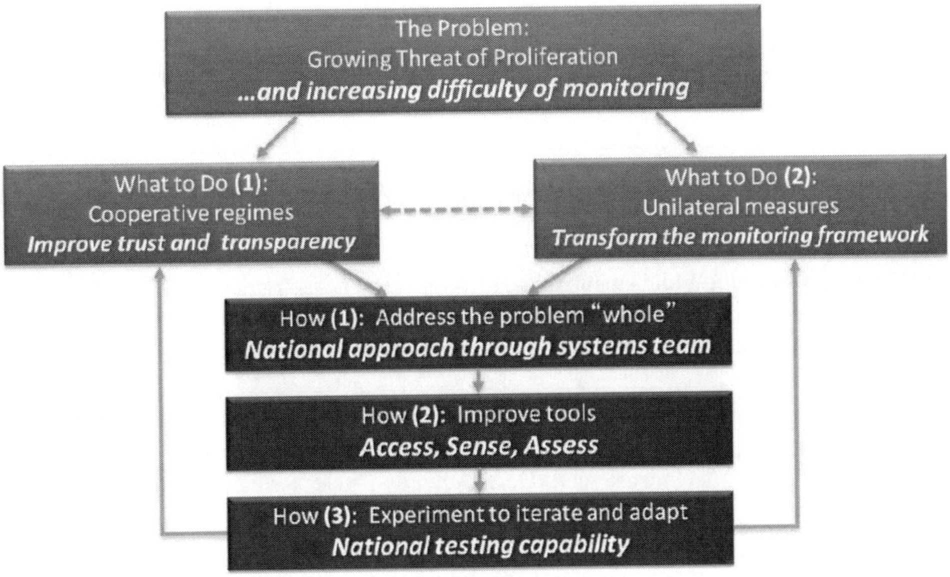

Figure ES-1. Plan for Long-Term Engagement and Commitment.

The technical approach for monitoring cannot continue to derive only from treaty and agreement dictates for "point" compliance to the numbers and types formally agreed upon and geographically bounded. Proliferation in this future context is a continuous process for which persistent surveillance, tailored to the environment of concern, is needed. This leads to the need for a paradigm shift in which the boundaries are blurred between monitoring for compliance and monitoring for proliferation, between cooperative and unilateral measures. Monitoring will need to be continuous, adaptive, and continuously tested for its effectiveness against an array of differing, creative and adaptive proliferators.

A Comprehensive Monitoring Framework: Two "Whats" and Three "Hows"

The Task Force observed early in its deliberations that there are many communities involved in tackling a piece of the monitoring "elephant," but found no group that could clearly articulate the entire problem nor a strategy for addressing it in any complete or comprehensive fashion.

The Task Force therefore created its own framework, illustrated in Figure ES-1, as a vehicle for communicating its findings and recommendations, but also as a potential starting point for the many agencies involved to see how their efforts might integrate into a more effective whole.

The framework consists of five elements – two "whats" and three "hows." Each of the five, as noted by the byline for each in the figure, has a top-level strategic objective (e.g., the strategic objective for Cooperative Regimes is to improve trust and transparency). The "hows" are intended to contribute to both "whats," and the overall architecture is built with a continuous cycle in which fielded results from the "whats" inform priorities and gaps in the "hows." In turn, new capabilities from the "hows" improve execution of the "whats." The proliferation problem and framework development, plus each of the five elements of the framework are covered in separate Sections of this report. The summary of the findings and recommendations for each of the five elements follows.

Cooperative Regimes: Key Findings and Recommendations

International Cooperation and Transparency

Principal Finding: While many relevant programs and activities exist throughout the government, the U.S. lacks a cohesive, long term, international engagement plan aimed at building cooperation and transparency.

Recommendation: State/Bureau of Arms Control, Verification and Compliance (AVC) (diplomatic), DOE/National Nuclear Security Agency (NNSA)/ Defense Nuclear Non-Proliferation (NA-20) (technical), and DoD/Defense Threat Reduction Agency (DTRA) (operational) should jointly develop such a plan and cooperatively implement their respective/shared responsibilities. The Task Force recommends a 4-phase approach for expanded cooperation, each step of which would gradually evolve to "internationalize transparency inspections." The Task Force approach could be enabled by expansion of the role of the International Atomic Energy Agency (IAEA) for assuming responsibility for the transparency responsibilities that ultimately emerge:

1. Bilateral, cooperative developments and evaluations among P-5 states, building on experience in historic allied partnership programs, as well as the Cooperative Threat Reduction (CTR) program with Russia;
2. Extension to all nuclear weapon states;
3. Expansion to nuclear materials transparency among major states with nuclear power generation;
4. Negotiation of a future Non-Proliferation Treaty (NPT "X") to bring in all nuclear weapon and material programs into a cooperative, multi-lateral regime.

The three lead offices cited should establish a Multi-Agency Roadmap in partnership with other relevant players in the government. The process should include the appointment of a (U.S.) Mission Manager[1] to drive the realization and coordination for this initiative throughout the development process.

Research and Development

Key Finding: Progress in building greater cooperation and transparency will require trusted technical support systems that do not currently exist.

Recommendation: DOE/NNSA/NA-20 should pursue an international research and development (R&D) program in automated monitoring and reporting systems supported by information barriers and authentication to enable more effective and extensive materials monitoring. Trusted information barriers, capabilities for real-time process monitoring and in-field inspection and analysis capabilities should be developed.

DoD/Assistant Secretary of Defense for Nuclear, Chemical, and Biological Defense Programs (ASD(NCB)), DTRA, and DOE/NNSA/NA-10 should partner to develop cooperative options for asymmetric nuclear weapons security paradigms; e.g.,

- Sharable technical security principles, practices, and technologies for sites, materials, and components
- Site declaration and portal monitoring
- "Assurance Volume" concepts

U.S Modernization Programs

Key Finding: DoD and DOE are starting to invest heavily in nuclear offense force and facility modernization. The Task Force found only limited consideration being given to more intrusive inspection regimes expected in future treaties and agreements, as part of the new design activities in each Department.

Recommendation: State/AVC, DOE/NNSA, and DoD/Acquisition, Technology, and Logistics (AT&L) should review current U.S. facility and weapon system modernization programs and instruct program managers, if necessary, to plan for accommodation of greater transparency measures. Methods for red teaming and performing vulnerability assessments should be developed and exercised routinely, for both existing and planned facilities, systems, and operations.

Unilateral Measures: Key Findings and Recommendations

Principal Finding: The guiding principle for monitoring to detect undesirable nuclear activity should be detection of activities as early in the planning and acquisition of a capability as possible in order to provide the greatest number of options for slowing or reversing the effort. New intelligence, surveillance, and reconnaissance (ISR) technologies, demonstrated in recent conflicts, offer significant promise for monitoring undesirable nuclear activity throughout the world. The nature of these technologies in the context of the monitoring challenge, however, is that the technologies are most effective when applied in an integrated architecture.

Recommendations: The IC (led by the National Counterproliferation Center [NCPC]), DOE/NNSA, DoD/DTRA and Department of Homeland Security (DHS)/Domestic Nuclear Detection Office (DNDO), should develop a joint roadmap, supported by the necessary systems analysis and engineering capabilities[2], to implement an integrated, more comprehensive and responsive monitoring architecture for nuclear weapons activities worldwide, expanding upon the more general, but static approaches currently employed. The roadmap should make clear the lead and support roles for all involved. As the legislated interagency lead for the Global Nuclear Detection Architecture (GNDA), DNDO should incorporate this more expansive monitoring architecture into the GNDA. The monitoring architecture should be structured to:

- Create "corridors of observation" in multiple domains (geographic, commercial, individuals, financial…) based on open source and available multi-intelligence (INT) information;
- Establish "patterns of life" within corridors that are suspicious;
- Focus persistent monitoring assets on individuals and activities of greatest concern that emerge from patterns of life analyses;
- Assemble and analyze data from all sources to support verification or identify previously unknown concerns;
- Use results of analyses to iterate—i.e., to provide ever more focus for intelligence taskings, as well as provide guidance for needed improvements in technical and operational capabilities.

National Security Staff (NSS) should request that DoD integrate the architectural elements into a global awareness system that:

- Builds on lessons and experiences of successful national security capabilities, such as the Counter IED Operations Center (COIC), NSA's counter-terrorism capabilities, NCPC's counterproliferation efforts, and on DOE's Proliferation Risk Analysis Program;
- Builds also upon the lessons and experience of the IC's Treaty Monitoring Manager3, which at the height of arms control monitoring in the late '80s and early '90s, proved to be an effective vehicle for coordinating the work of multiple organizations and collection activities;
- Uses the Monitoring and Verification (M&V) National Testing Capability (described and recommended in Section 6) as the key operational assessment vehicle to test a range of scenarios.

The IC should recast DOE/IN's historic "nth country" analytical capabilities into an "nth group" effort, adapted to the wider range of actors and designs accompanying both vertical and horizontal proliferation.

- The "nth group" program should serve as a community asset for helping to characterize proliferation pathways from available data and advancing technology, and posit pathways to guide collection priorities.

Address the Problem "Whole": Key Findings and Recommendations

Principal Findings: Experienced professionals in M&V have declined in number, but those engaged still tend to address the problem as an extension of approaches used in past treaties and agreements. As further agreements drive both the U.S. and other signatories to smaller numbers of weapons, the price for inaccuracies increases. The monitoring challenge is further complicated by the many dimensions of proliferation discussed above.

A more robust approach is needed, one that is derived from common community characterization of the problem in its full breadth, depth, and extent; i.e., a systems approach.

In addition, there is a need to re-grow a knowledgeable workforce, especially systems analysts to support threat assessment, trade-off analyses, and investment prioritization. Historic contributors such as the DOE national laboratories remain critical to addressing the nonproliferation problem. However, the Task Force believes that the responsible agencies would benefit from enlisting a wider array of performers (other labs, contractors, academia) to address the full complexity of this problem space that calls for new, as well as improved, tools.

Recommendations: As the lead agencies, State/AVC, ASD(NCB), NNSA and NCPC should create the processes and oversee the following steps:

- Establish a "White Team" whose charter is to characterize the comprehensive Monitoring and Verification Framework, relating threat events and actions to monitoring requirements, both cooperative and unilateral.
    - The White Team should be assigned to a "home" agency, but supported and governed by an interagency "board" whose members have sufficient authority to influence any needed changes in strategy and program directions at their home agencies.
- Ensure a common understanding among agency leads for addressing all aspects of the framework, including policy, diplomacy, operations, and research, development, test & evaluation (RDT&E).
    - It is critical that those agencies involved in implementing M&V capabilities maintain a high degree of unanimity on how M&V problems and challenges are characterized, and how they will be addressed in order for the White Team to be both effective and sustainable. The adoption of common frameworks (such as that proposed in this report) can contribute to developing a common understanding of national strategy, goals, and pathways for accomplishing those goals.

The foundation established with the two steps above leads to the following implementation steps, again under the charge of the leading agencies:

- Adapt or create integrated implementation plans. No single agency has purview over the totality of the responsibilities in the M&V mission, but integration "across the seams" is required.
    - The NSS, with support from the Office of Science & Technology Policy (OSTP), should assess progress annually.
- Establish an institutionalized interagency planning process that evolves with the threat.
    - The nature of the M&V problem and solution spaces has changed considerably over the past decades. Undoubtedly, the challenges and proposed solutions will continue to evolve in the coming years. Therefore, not only must the analytical framework used by the White Team evolve, but the plans and priorities across the government must evolve as well.
- Enlist a wide range of contributors.

Improve the Tools: Key Findings and Recommendations

Improving Access

Principal Finding: Improving access is essential for making progress against the low and/or obscured signal of incipient proliferators.

Recommendation: The responsible agencies in the IC and DoD should rebalance existing investments and/or grow new programs in R&D to develop improved approaches for obtaining access across an array of scenarios and environments.

Expanded Sensing and Assessment

Principal Findings: The advances in persistent surveillance, automated tracking, rapid analyses of large and multi-source data sets, and open source analyses to support conventional warfighting and counterterrorism have not yet been exploited by the nuclear monitoring community. Conversely, developers of these capabilities are largely unaware of the challenges and requirements for nuclear activity monitoring.

Radiation detection remains important for monitoring and verifying sources of special nuclear material (SNM) when access allows inherent range and background limitations to be overcome. The technical communities involved have had a mixed track record of fielding new capabilities in spite of coordinated and sustained investments in R&D. The problem appears to lie in the gap between researchers and operators, who do not come together early enough in the development process to ensure that both technical and operational requirements are addressed.

Two specialized capabilities critical to monitoring require support to enable the monitoring architecture proposed in this report:

- The Open Source Center (OSC), which is proving to be a particularly effective organization in helping to identify the "corridors of interest", but has limited resources to devote to routine monitoring of other groups or nations.
- The U.S. Atomic Energy Detection System (USAEDS, operated by the Air Force Technical Applications Center), which remains a stepchild within the Air Force (AF) and is in serious need of modernization.

Recommendations: There are numerous aspects that deserve attention from each or all of the several agencies with mission responsibilities for nuclear monitoring, as follows:

- The Director of National Intelligence (DNI) should provide direction to the leading agencies in the IC to:
    - Increase the profile, support, and integration for the OSC's Counterproliferation (CP) Program to collect and disseminate information and analysis relevant to arms control and proliferation issues;
    - Assess/adapt new and expanded collection capabilities for nuclear monitoring, especially through multi-INT integration and enhanced processing;

- Expand the use of open source and commercial information to focus search areas and reduce demand on national collection assets so that the collection system can keep up better with the expansion of targeted areas of interest;
    - Continue/expand the augmentation of data from national intelligence collection systems with imagery and radar from commercial systems;
    - Develop and apply quality assurance methods for crowd sourcing of commercial imagery results;
- Ensure that DoD's activities for improving global indications and warning leverage existing sources of information and capabilities, and develop the analytics to produce actionable nuclear-related threat warnings;
- Adapt the Treaty Monitoring Manager's role to one of a Proliferation Monitoring Manager, "home based" in NCPC, with orchestration and integration responsibilities to assess both horizontal and vertical proliferation.

- DOE/NNSA, DHS/DNDO, DoD/DTRA and the IC should build upon the existing memorandum of understanding (MOU) among these organizations to improve coordination and execution of their respective radiation detection programs to:
    - Conduct systems studies and engage operators early in development to improve transition of radiation detection advances to the field;
    - Ensure that developers and users agree in advance on system concepts, measures of success and levels of readiness for the principal technologies and operational scenarios of interest;
    - Focus new efforts on accelerating development of research with near-term payoffs, and investment in longer range technologies that can meet both technical *and* operational feasibility requirements;
    - Develop managed access to nuclear facilities and test ranges by all involved agencies at which detection technologies and operational approaches can be explored using real SNM;
    - Agree on investment strategies. There should be supporting inter-agency roadmaps to integrate efforts and focus crosstalk. Annual inter-agency reports to the NSS should be issued to track progress and enhance information flow.[4]
- DOE/NNSA, DTRA, and the IC should rebalance existing investments in order to grow new programs in R&D that expand activities and the supplier base to include adaptation of conventional warfighting ISR advances: e.g.,
    - Engage in planning and capabilities development, especially for data collection and fusion functions, for DoD's efforts to improve nuclear situational awareness;
    - Support transition of multi-INT fusion and exploitation tools to nuclear monitoring applications;
    - Ensure activities related to nuclear weapons and materials monitoring are guided by the "white team" function as discussed in the previous section and Section 4;
    - Make explicit the requirements for, and improvements needed in, HUMINT, SIGINT, cyber, OSINT, etc., to support monitoring and verification.
- NSS should monitor closely and persistently the resourcing to modernize the USAEDS with the help of the Vice Chairman of the Joint Chiefs of Staff (VCJCS), who should ensure its modernization, is supported in AF budgets.

Test to Iterate and Adapt: Key Findings and Recommendations

Findings: The comprehensive monitoring regime the Task Force proposes is a system of systems that must work together. It is too complicated to plan or assess analytically or with piecemeal testing, especially when used against an adaptive adversary and/or one with sophisticated denial and deception.

Furthermore, experience shows that operators typically learn to use fielded systems differently and often better than system design/analysis would predict—an effect that would result from into iterative development and training.

Recommendations: DTRA, in partnership with DOE/NNSA and the IC, should develop comprehensive "iterate and adapt" national testing capabilities which:

- Provide a focal point for planning, iterating/adapting, and operating the system of systems;
- Help integrate technically disparate and organizationally disaggregated activities.

The national testing capabilities should include four interdependent parts:

- Ranges and facilities (real and simulated), almost all of which currently exist;
- Information/data management systems (e.g. data flows from and to sensors, information flows within analysis/fusion systems, and command/control data/information flows);
- A standing/ongoing Red Teaming activity;
- A White Team—a strong analytical component to assess the implications of test results and feedback into the cycle needed changes/improvements.

Support from the IC (principally NCPC and DOE) should be provided for the purpose of rebuilding a national nuclear threat technical assessment capability that anticipates "nth group" designs, effects, employment and deployment approaches and observables (per one of the recommendations for "Unilateral Measures").

Summary

The problem of managing the global nuclear environment to maintain stability will be with us for a very long time.

The best bet for making progress is a sustained effort in which experience, competence, and relationships can be built over successive careers. The challenges are daunting and success will be difficult to come by, but prior experience tells us that persistence can lead to the outcomes we seek.

SECTION 1. THE PROBLEM – GROWING THREAT OF PROLIFERATION AND INCREASING DIFFICULTY OF MONITORING

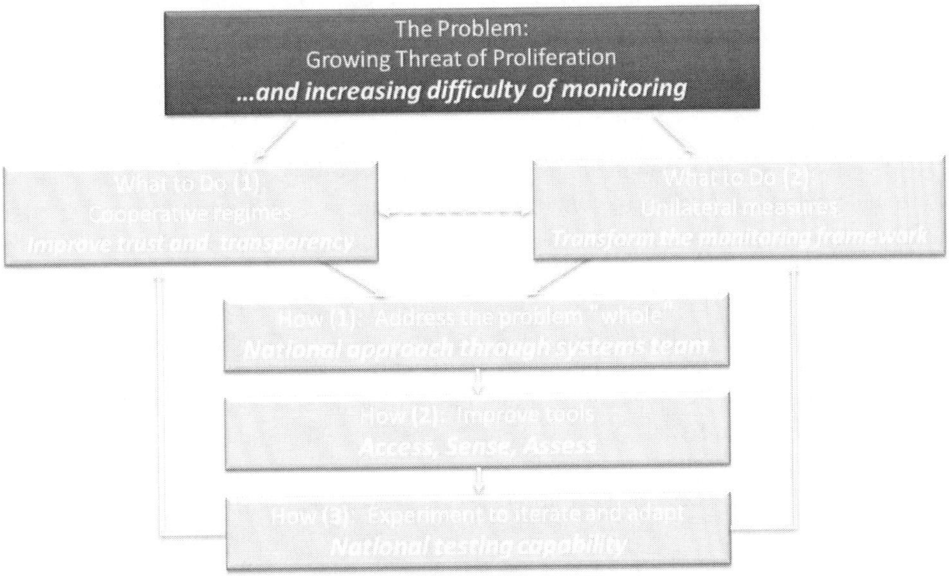

1.1. Monitoring vs. Verification

The Terms of Reference—indeed the very name of the study—refer to monitoring and verification technologies. The Task Force focused, however, only on capabilities for monitoring because verification, traditionally and in the view of the Task Force, is principally the political judgment to which monitoring and other means contribute. Verification is also most often used in the context of adherence to, or violations of, treaties and/or formal or informal agreements. Monitoring, however, can and should be more broadly applied because of the nuclear future envisioned by the Task Force. In fact, monitoring for proliferation should be a top national security objective – and one that the nation is not yet organized or fully equipped to address.

1.2. The Nuclear Future – Not an Extrapolation of the Past

Too many factors have changed, and are changing, from our historic basis and experience in the Cold War, in a manner that should give national leadership pause for concern. The list starts with the fact that fundamental nuclear knowledge is much more widespread. Ubiquitous information access and widespread observational tools are increasing inherent transparency. At the same time, recognition of such increased transparency by potential or actual proliferants naturally leads to more sophisticated methods of denial and deception.

The actual or threatened acquisition of nuclear weapons by more actors, for a range of different reasons, is emerging in numbers not seen since the first two decades of the Cold

War. Many of these actors are hostile to the U.S. and its allies, and they do not appear to be bound by established norms nor deterred by traditional means. In some cases of established nuclear powers, nuclear forces are seen as the most affordable and effective alternative to deter superior conventional forces; i.e., nuclear weapons are viewed as a legitimate warfighting capability, especially if their vital domestic or regional security interests are threatened. For example, Russia has publicly stated in doctrine and backed it up with training and exercises that they will use theater nuclear forces if necessary to deter aggression against the homeland.[5,6,7,8,9]

The pathways to proliferation are expanding. Networks of cooperation among countries that would otherwise have little reason to do so, such as the A.Q. Khan network or the Syria-North Korea and Iran-North Korea collaborations, cannot be considered as isolated events. Moreover, the growth in nuclear power worldwide offers more opportunity for "leakage" and/or hiding small programs, especially since current resources to support safeguards are already strained and will be increasingly challenged by cases of noncompliance.

In short, for the first time since the early decades of the nuclear era, the nation needs to be equally concerned about both "vertical" proliferation (the increase in capabilities of existing nuclear states) and "horizontal" proliferation (an increase in the number of states and non-state actors possessing or attempting to possess nuclear weapons).

The challenges for monitoring in this context are much more difficult. Historically, and even with New START, monitoring has focused on relatively few nations (only two in treaties with Russia) and locations. Moreover, the objects to be monitored have been numerous and easily identifiable (e.g., delivery platforms such as bombers, missiles, and submarines), the facilities supporting the enterprise visible and often declared, and nuclear materials inventories voluntarily declared.

In the nuclear future as seen by the Task Force, monitoring will need to address more widespread foreign nuclear weapons related activities in a "messy" combination of negotiated, non-cooperative, and non-permissive environments. Any or all of the following could be factors in monitoring a particular nation or group of concern:

- Small inventories of weapons and materials, even as low as a single "significant quantity of fissile material;"
- Small nuclear enterprises designed to produce, store, and deploy only a small number of weapons – either as an end goal, or as the first steps of a proliferant, or a nuclear terrorist operation;
- Undeclared facilities and/or covert operations such as testing below detection thresholds;
- Use of non-traditional technologies, presenting at best ambiguous signatures, to acquire both materials and components;
- Theater nuclear forces and associated doctrine, exercises, and training complicated by the use of mobile, dual use delivery systems;
- Many more players to whom access by the U.S. or its allies will be limited or extremely difficult, some of whom will be globally networked with global access to relevant science and technology.

The stress on monitoring technologies is significant. Identifying and maintaining track on people, nuclear components and warheads now becomes a requirement, and in many cases, physical access will be limited. Moreover, the broad access to technology and the growing sophistication of cyber offenses by "them" as well as us will allow others to gain insights into our collection capabilities and methods. The challenges are further compounded by the many dimensions of the nuclear enterprise that need to be monitored, as illustrated in Figure 1.1.

The materials themselves in their many chemical forms, the operational configurations to which the materials are applied and associated facilities and infrastructure are all part of the enterprise that should be monitored. Nuclear energy operations, as well as nuclear weapons activities, should be carefully watched because of the potential for diversion of materials from energy to weapons, the ease of hiding low level weaponization activities within nominally civilian facilities, and/or the requirement for much lower quantities of weapons grade vs. reactor grade materials to obtain an operational weapons capability.

1.3. A Need for Comprehensive Monitoring

Given the potential reality of, and subsequent risks associated with, the nuclear future described above, the Task Force believes that the nation must "address the proliferation problem whole" because the global nuclear regime is itself an integral whole. Civil nuclear applications are a global enterprise, and military nuclear applications are increasingly networked. Military and civilian applications are connected, in part through the global nuclear science and technology (S&T) base. Actual or latent proliferation by one nation stimulates proliferation activities by others, especially neighbors.

	Nuclear Energy	Nuclear Weapons
Materials	▪ Pu, HEU, spent fuel - Metallic (pure and alloyed), oxide w/ highly varying densities - Transmutation - Quantities of tons	▪ Pu, HEU, high level waste - Civilian complexities plus... - Secrecy of isotopic composition - Quantities of kg's
Facilities, Operations	▪ Enrichment and fabrication ▪ Reprocessing ▪ Storage - materials ▪ Reactors ▪ Laboratories ▪ Transportation...	▪ Enrichment and fabrication ▪ Storage - materials and weapons ▪ Laboratories ▪ Transportation ▪ Deployed weapons ▪ Training and exercises...

Diversion

Figure 1-1. Challenges are Further Compounded by What Needs to be Monitored.

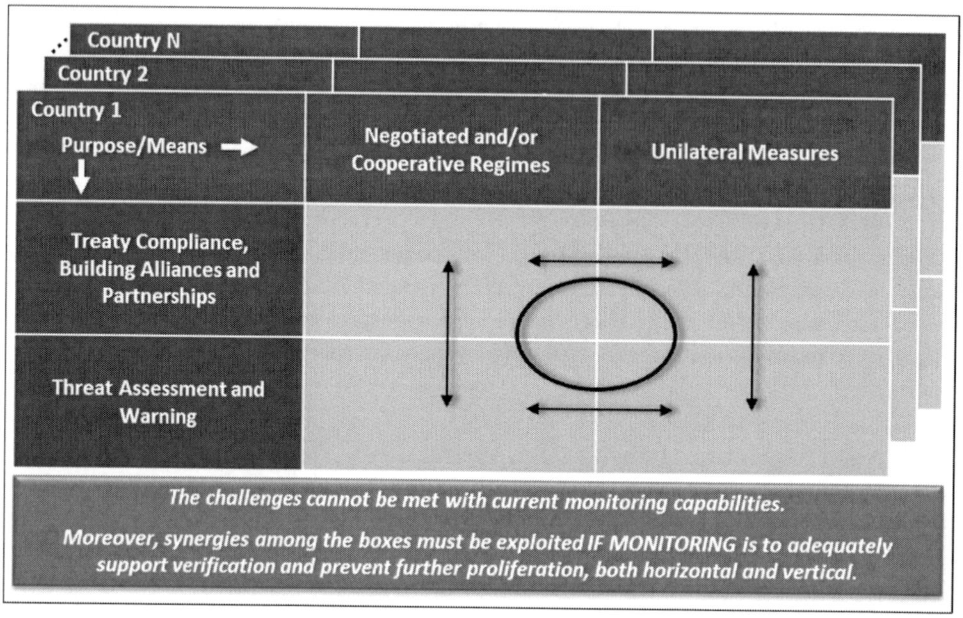

Figure 1-2. Scope – and Opportunities – of Future Monitoring Regimes.

In order to understand any one part of such a global interconnected enterprise (for example, to negotiate and/or monitor a particular treaty, or understand the threat posed by a nation that has proliferated), it is necessary to understand the whole. Understanding the whole entails monitoring the whole, and not all of that monitoring can be carried out through methods negotiated in treaties, at least not until the global treaty regime has become much more comprehensive. Negotiated monitoring for a particular treaty will have to be complemented by non-negotiated, "general purpose" monitoring, including unilateral monitoring. The Task Force acknowledges that such a relationship between negotiated and non-negotiated monitoring has been important in the past, but it will be even more important in the future. Moreover, the details (e.g., geography, access, suspected stage of proliferation, etc.) of where, and to what extent, each are applied will be important; one size will not fit all.

Figure 1-2 illustrates the point. Not only does the figure highlight the expanded and multi-dimensional nature of the proliferation monitoring problem, but it also introduces the opportunities for *synergy* among different types of monitoring for different purposes.

Managed as a whole, synergies can be developed in both the horizontal and vertical dimensions of the purpose vs. means matrix. Horizontally, for example, non-compliance with the provisions of a treaty might involve hiding assets or undertaking covert operations that non-negotiated/unilateral monitoring can uncover. Challenge inspections negotiated in a treaty can be triggered by non-negotiated monitoring data. Similarly, "general-purpose" monitoring for threat assessment (lower right) can be complemented by understanding treaty-limited items or behaviors learned through negotiated monitoring (lower left).

Vertically in Figure 1-2, an overall assessment of threats of proliferation or threats by proliferators constitutes the context in which future treaties should be proposed and negotiated. Conversely, observed non-compliance with a treaty can serve as a form of strategic warning in the threat assessment sense. As a more particular example, monitoring for "patterns of life" of nuclear operations in a particular nation (lower right) can indicate or reveal the presence of evasive/covert operations intended for non-compliance with treaties (upper right, moving to upper left).

Figure 1-2 also illustrates a separate monitoring approach for each of several nations to indicate, as discussed above, that nuclear activities can be both unique to each and increasingly networked. While complicating the overall challenge, this characteristic introduces more opportunities for synergies by integrating monitoring across several/all nations.[10]

1.4. Beyond Incremental Improvements… A Paradigm Shift

Because of the increasing challenges of controlling, limiting, and stabilizing the global nuclear regime and the increasing difficulty of monitoring it, the Task Force believes that *efforts beyond incremental improvements of traditional approaches to monitoring will be required.* Monitoring cannot continue to derive only from treaty and agreement dictates for "point" compliance to the numbers and types formally agreed upon and geographically bounded. Proliferation should become a continuous process for which persistent surveillance tailored to the environment of concern is needed. This leads to the need for a paradigm shift in which the boundaries are blurred between monitoring for compliance and monitoring for proliferation, between cooperative and unilateral measures. Monitoring will need to be continuous, adaptive, and continuously tested for its effectiveness against an array of differing, creative and adaptive proliferators.

The Task Force envisions a five-part framework, illustrated in Figure 1-3.

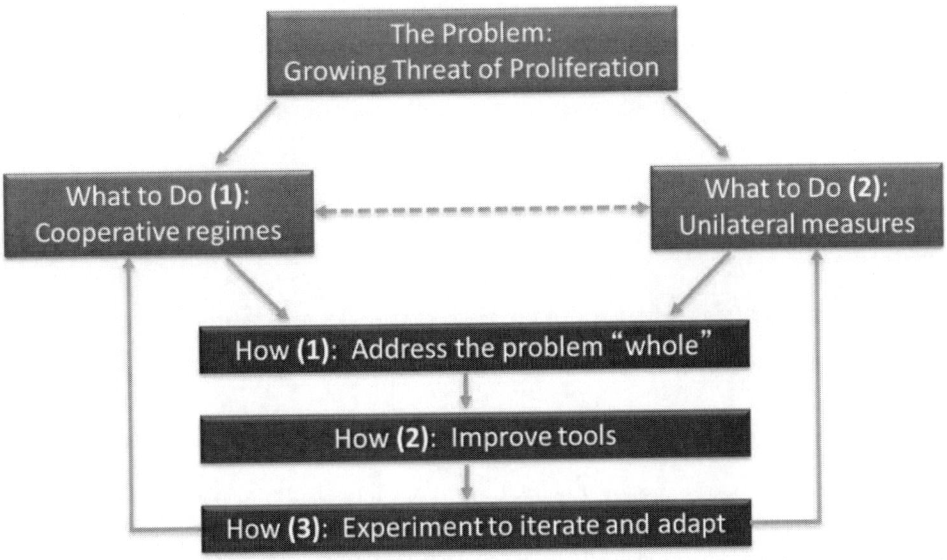

Figure 1-3. Meeting the Future Monitoring Challenge: Two "Whats" and Three "Hows".

Figure 1-3 highlights that it will be essential to expand the scope of monitoring by achieving greater international cooperation and transparency, and by developing and employing new U.S. monitoring architectures. Both will require that the monitoring problem be approached systematically and "whole;" improved through technical tools that enable access, sensing, and assessment; and continuously tested against evolving and adaptive proliferant strategies and techniques.

To achieve these goals, a major diplomatic effort to build trust and confidence will be required. Investments will also be needed to develop and field capabilities beyond the traditional emphases in treaty-monitoring technologies. Two key examples illustrate the point. First, the powerful tactical ISR developed for Iraq and Afghanistan – for example, for suppressing the improvised explosive device (IED) threat – should be adapted and extended for nuclear monitoring, especially where access is limited or denied. Such capabilities could also be negotiated for use in challenge inspections. Second, the "big data" technologies for extracting meaning from vast quantities of data that are being developed commercially in the information technology (IT) industry, and for other purposes in DoD and the IC, need to be extended and applied to nuclear monitoring.[11]

Nations seeking to proliferate and/or to evade the provisions of treaties (as well as non-nation state adversaries with nuclear ambitions) will be adaptive in hiding or obscuring what they are doing. Staying ahead of their adaptation must be an integral and deliberate part of U.S. efforts to develop and implement a comprehensive nuclear monitoring regime. A key element of staying ahead is continuously challenging our own assumptions. The Task Force believes this could best be accomplished through testing and experimentation in which monitoring capabilities would be challenged by red-teaming, with the red/blue interactions analyzed and refereed by a "white team." In this way, we can "try before we buy", and account for others' use of technology to thwart our own.[12]

In summary, the challenges of controlling and stabilizing the nuclear future, and the difficulty of monitoring global nuclear activities in that future, mean that the nation must plan for a long period of building both the political *and* technical groundwork for the next major steps in formal treaties or agreements, as well as for addressing proliferation more broadly where cooperation is unlikely for the foreseeable future. To drive the point home, the Task Force adopted the motto: "We can't let our treaties get ahead of our monitoring and verification headlights."

SECTION 2. COOPERATIVE REGIMES: IMPROVING TRUST AND TRANSPARENCY

2.1. Introduction

There are many trends already at work toward openness and transparency in technology and in global institutions in general, in spite of resistance to and/or attempts to limit them in selected quarters. International trade and transportation, multinational corporations, international contract law, treaties intended to manage resources in the global commons, the Internet and the growing ubiquity of communications and information access are all cases in point. Each of these domains has largely developed its norms, standards, and/or policies independently and according to its own needs.

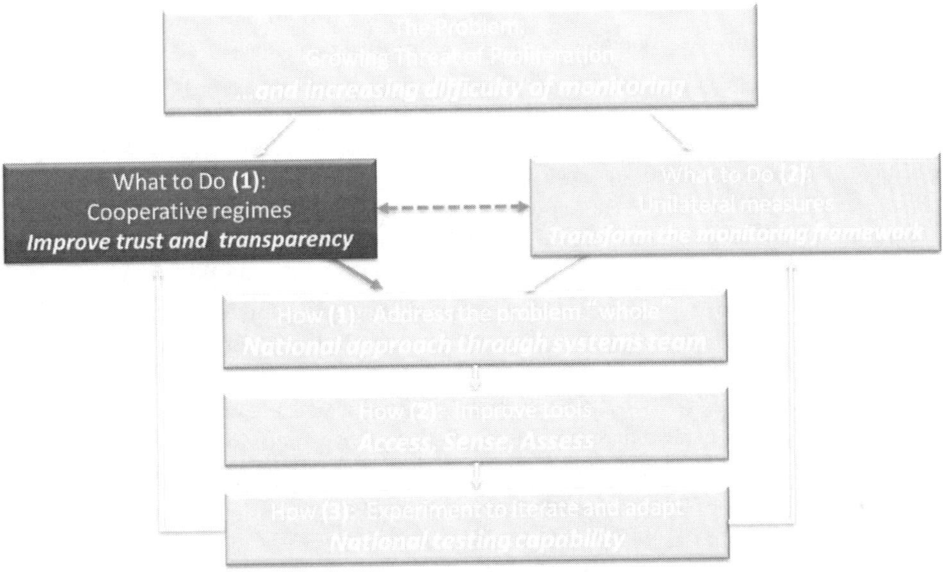

The increase in openness and transparency in this wide range of domains provides a foundation to build upon, extend, and integrate to raise the levels of trust among participants––most especially for monitoring proliferation related activities. However, because nuclear weapons are so central in security relations among nations, achieving higher levels of transparency and access needed for effective monitoring will require a comprehensive, sustained, policy-based diplomatic approach coordinated across the U.S. Government (USG) and with other nations, devoted expressly to advance the cause of openness and transparency writ large. While the U.S. has recognized the value of, and need for, such an approach almost since the development of nuclear weapons, it has had, at best, a mixed record of success in sustaining national efforts. This situation should be addressed with the highest priority, and the Task Force provides in this Section one approach for how to do so.

2.2. A Well Established Starting Point: The Nonproliferation Treaty and Its Limitations

A baseline of nuclear transparency has been established with the Treaty on the Non-Proliferation of Nuclear Weapons (NPT).[13] However, the NPT regime remains under significant pressure—both from the expansion of nuclear power with the associated monitoring demands, and the continued intransigence of states violating their safeguard obligations.

In order to diversify their energy portfolios and meet growing demands for electricity, new states are venturing into civilian nuclear power. Most, if not all, of this expansion will be benign, but the dual-use nature of nuclear technology, at least with regard to SNM production, raises a number of concerns. Some of this growth will occur in areas of instability and/or regional tension. The recent events of the "Arab Spring" illustrate the potential for political volatility and range of possible outcomes.

Sensitive nuclear technologies, particularly those used in enrichment and reprocessing facilities, present unique monitoring and verification challenges. There are inherent difficulties monitoring complex, industrial size bulk handling facilities and processes. Expansion of nuclear power risks the illicit spread of these capabilities. The Kahn network illustrated the difficulties of controlling key technologies in a globalized and increasingly sophisticated manufacturing base. Collusion amongst proliferators enables states to bypass technological hurdles, serving to further complicate nonproliferation efforts.

In spite of these complexities, it is still decidedly easier to monitor declared material and facilities as the NPT (further enabled more recently by the Additional Protocol) has enabled for 40+ years, so that expanding the fraction of nuclear activities under "routine" inspection is desirable. Cooperative monitoring regimes provide a baseline of information while defining legitimate nuclear behavior. In recent years, linking nuclear security with Safeguards is providing a basis for engagement and cooperation between states to promote high standards for material protection, control and accounting. Cooperative monitoring also allows focusing of resources on "rest of world" coverage if/when concerns arise. International or multilateral verification activities provide independent data to support intervention efforts in cases of noncompliance.

The Safeguards system of the IAEA will need added resources and technical advances to deal with the effects of a global nuclear energy expansion to ensure a robust detection capability for any misuse. Should future arms reductions or other regimes or initiatives, such as Global Nuclear Lockdown and/or a Fissile Material Cutoff Treaty, require the IAEA to monitor a much more extensive set of facilities and/or materials, the existing Safeguards regime will have to shoulder an increase in the demand for strong but efficient monitoring and verification capabilities. Providing confidence that treaty obligations are being fulfilled while protecting sensitive information is a technical challenge common to both warhead verification and safeguards at commercial facilities, and one for which there are not yet widely accepted solutions.

2.3. Experience in Cooperation Outside the NPT: The Cooperative Threat Reduction Program

The CTR program was begun in 1993 in order to assist the Russians and the republics of the former Soviet Union to reduce the threat of a nuclear weapon accident or loss and meet their commitments to the original Strategic Arms Reduction Treaty (START). Also known as the Nunn-Lugar program,[14] the CTR program permitted the Ukraine, Belorussia and Kazakhstan to fulfill agreements to return their nuclear weapons to Russia and become nuclear weapons free states, reducing by three the number of governments that possessed operational nuclear weapons.

The program also enabled Russia to meet its commitments to START. The United States, by partnering with the Russian Defense Ministry to destroy missiles, bombers and submarines, helped ensure that the Russians reduced the numbers of delivery vehicles agreed to under the START Treaty. The CTR program also built facilities and provided security systems to store nuclear materials and secure operational nuclear weapons.[15] Funding was also provided to support Russian nuclear scientists with technical work in order to mitigate the concern that they might otherwise sell their expertise to other nations who are intent on building their own nuclear weapons capability. Participants on both sides agree that greater transparency and stability were achieved between the United States and Russia through the execution of the Nunn-Lugar program.

2.4. Toward an International Cooperation and Transparency Regime

Currently the IAEA monitors all declared stocks of SNM in states that are signatories to the NPT and occasionally SNM offered up as excess material in weapon states. The bilateral Safeguards agreements that provide the legal basis for safeguards implementation, require that the information collected be considered sensitive and not be shared with other states.[16] Extending protocols to include the monitoring of nuclear weapons and weapon components worldwide coupled to sharing of the information with all international parties who have agreed to cooperate should be the ultimate goal. In order to get to that point, a long period of trust building through a number of intermediate steps will be required. The Task Force examined the experience base to understand how to get started in such a process and then developed a proposal for a phased approach.

The Task Force started with an examination of the recent history of proposals and negotiations for how best to obtain mutual agreements on M&V technologies in the future. One of the best historic examples was the Cooperative Safe, Secure Dismantlement (SSD) talks of 1991-1994 (bilateral discussions between the United States and the Russian Federation). The talks had made significant progress towards a joint agreement for radiation detection equipment that could be used to establish the fact that an object presented for dismantlement was in fact a nuclear warhead.

Although these talks were terminated without a formal agreement being consummated, there were enthusiastic beliefs on both sides that they had "come close" to developing radiation detectors and related instruments and procedures that could adequately determine the amounts of SNM [either Highly-Enriched Uranium (HEU) or Plutonium (Pu)] with sufficient accuracy to establish that there were indeed weapon-like quantities present. The remaining barrier to be overcome was to ensure that sensitive information (such as specific design details of the devices presented) could not be transferred.

The approach being pursued during the SSD talks focused on the use of potential "Information Barriers," such as templates provided by the inspected party on CDs or magnetic disks at the point of inspection, which would then be used to convert the measured data into kilogram amounts of SNM, but which *would not* reveal geometric internal details of the device. That approach still had some difficulties prior to its being accepted, but there were serious efforts being expended by both the United States and Russian participants to find acceptable solutions. Methods such as this one could be the first step to multilateral cooperation that could include non-nuclear weapons states.

Over the past decade, programs such as the United States-Russia Warhead Safety and Security Exchange (WSSX), and the United States-Russia-IAEA Tri-lateral Initiative have explored transparency methods and ways to protect sensitive information.

In a recent exchange meeting on this topic in January of 2011 between scientists of the Russian and United States National Academies, both sides expressed strong interest in finding ways to complete the search for mutual agreement on measurement devices that could overcome the difficulties listed above, convinced that these will be essential for verification of future nuclear agreements. Efforts are currently underway to pursue a United States-Russian agreement to allow such cooperative work to be reinvigorated. In addition, there was a meeting of United States and Russian nuclear weapons lab directors late in 2012, and both sides spoke favorably of finding a way to renew cooperative work, that was lost when the Lab-to-Lab efforts fell off during the Russian/Commonwealth of Independent States (CIS) reorganization.

During its deliberations, the Task Force learned there has been additional U.S. thinking and work that would utilize film recording, rather than direct counting apparatus. There are strong beliefs that the needed systems can be achieved by employing random scanning techniques which could average out geometrical details, while preserving the total assay/inventory of special nuclear material amounts.

The positive experience of these and other examples led members of the Task Force to conceive an overall approach, and to recommend that the U.S. pursue joint development efforts for "**Building Cooperation and Transparency**." The Task Force envisions a multi-year effort, which can pay large dividends in terms of a universal transparency that would improve strategic and tactical stability against nuclear war among all nuclear-weapons states, as well as achieve enhanced confidence building for nonproliferation efforts. We endorse these approaches, and outline some of the details for the proposed phased initiative below.

Realizing that there are no perfect transparency regimes, there are still many advantages that can result when nations can achieve full reciprocity in monitoring both nuclear weapons and nuclear materials that might otherwise be diverted for weapons use. All parties would benefit from the national security stability that would ensue from having transparent knowledge of the numbers/types of other nations' nuclear arsenals, while each nation in turn makes the knowledge of their own SNM and/or nuclear weapons inventories available to the others. The Task Force envisioned that the same techniques could be utilized by inspectors for routine inspections, dismantlement verification, and warhead destruction monitoring.

There are of course essential criteria that must be met before such a regime can be successful. Of primary importance, the security of storage sites and the security of stored weapons or materials should not be undermined or even weakened by the agreed upon inspection processes to be used. Similarly, successful transparency should reveal neither national secrets, such as specific locations of storage sites or security design techniques, nor vulnerabilities within the weapons designs. Achieving acceptance by all parties as to the permissible level of intrusiveness during the inspections, while preserving the effectiveness of the security measures employed by the inspected parties, will require independent agreement and confidence in these factors by all of the parties.

The Task Force fully understands that the task proposed will not be easy nor will it be accomplished soon. However, the Task Force does believe that the times are now propitious to move forward on a path to develop universal transparency regimes that can simultaneously fulfill these goals and requirements through an international process for achieving universal knowledge of nuclear weapon inventories and SNM inventories, and that the U.S. should lead in such an effort. Indeed the U.S. has already declassified the size of its current nuclear arsenal. The Task Force proposal has four phases, each with subparts that involve cooperative development efforts with a multiplicity of other states.

Phase 1. Bilateral Cooperative Developments/Evaluations with P-5 States

The journey should begin with a series of bilateral efforts among the five permanent members of the United Nations Security Council (P-5) nations (United States, Russia, UK, China, France) to jointly develop, evaluate, and improve monitoring equipment and recording methods to demonstrate the capabilities needed for warhead verification. Particular emphasis should be placed on achieving acceptable levels of "non-intrusiveness." In order to win agreement for use of any approach by all of these parties, it will be important to prove that sensitive information will not be revealed—either directly or through collateral or surreptitious means. The model of the Joint Verification Experiment (JVE) provides an excellent template for how similar bilateral cooperative efforts, which the Task Force envisions here, could proceed. The JVE was carried out cooperatively between the United States and the U.S.S.R. in 1987 and 1988 for the purpose of creating mutually agreed upon methods for both on-site measurements and on-site verification systems in support of the series of bilateral Nuclear Testing Talks.[17]

When successfully demonstrated and accepted by the parties, these monitoring systems could be used for verification of nuclear weapons treaties, for authentication of declared stockpiles, for verification of dismantlement, and ultimately for verification of the destructions of warheads. At some point, the bilateral nature of this phase should move to multilateral cooperation across the P-5.

Phase 2. Extending the Development/Evaluations to All Nuclear Weapon States

The P-5 would embark on cooperative, bilateral efforts to demonstrate, adapt, and improve the systems through joint development efforts with all nations that currently possess any nuclear weapons. This would be a very sensitive endeavor, as previously there have been few if any contacts made with these nations (e.g., India, Pakistan) regarding measurements or monitoring of their nuclear weapons. The goal would be the same as for the previous phase: to demonstrate accurate and non-intrusive monitoring in order to win acceptance for use of demonstrated P-5 techniques or variants by all of these states. During this phase there will need to be a premium placed on the flexibility of implementing the monitoring techniques, since there are large variations in security and control measures, technical sophistication, and inspection methodologies. Initiation of this phase need not wait until Phase 1 is completed since this effort will likely require a lengthy period to establish mutual understanding and trust well before technology options or even discussed or tested.

Phase 3. Major Nations That Employ Nuclear Power Generation

The next phase shifts from nuclear weapons transparency to engaging Non-Nuclear Weapons States (NNWS). The particular focus should be on adapting existing international safeguards or new transparency techniques to achieve accurate and sharable determinations of the inventories of separated SNM (or weapons useable materials) in such states as Japan, Belgium, Netherlands, Germany, Canada, Taiwan, Ukraine, Kazakhstan, South Africa, and others who currently have or expect to generate in the future any separated quantities of material. Each nation must also believe that the techniques will be accurate, reciprocal, and non-intrusive in ways that would protect its commercial secrets, and would not compromise the security for storage of any of its materials. The efforts would be cooperative and bi- or (better) multi-lateral.

It is in this phase that the emphasis for Building Global Transparency of Nuclear Materials would become the dominant theme. At present, the IAEA retains the sensitive information as it carries out the verification activities on behalf of the member states. The shift from current protocols and practices under the NPT for such a Global Transparency regime would be making transparency information available as a quid pro quo for providing access to an international inspectorate.

Phase 4. The Evolution from Bi-lateral to Multilateral Implementation and a Prospective Non-Proliferation Treaty (NPT-x)

Several studies have concluded that it is premature to pursue negotiation of a follow-on Non-Proliferation Treaty that would impose transparency on States possessing nuclear weapons and NNWS equally and add nuclear weapons disarmament transparency to the treaty. Completion of the three phases presented above, however, could set the stage for overcoming the current difficulties and be the basis for the trust and understanding needed to carry out both the periodic/continual monitoring of nuclear weapons worldwide, and the periodic/continual monitoring of SNM quantities of potential nuclear weapons materials worldwide.

The signatories of nuclear arms control or arms reductions agreements, joined by all of the nuclear weapons-possessing nations, would collectively and mutually negotiate the procedures, frequencies, prohibitions, etc. for carrying out materials and weapons transparency measures/inspections protecting against the spread of nuclear weapons expertise to NNWS. The ideal outcome would be agreement that the results of these inspections would be delivered to the IAEA as part of its routine monitoring and shared with all nations worldwide. The Task Force believes that progress through Phase 4 will have a positive effect on worldwide arms stability as well as strengthen non-proliferation efforts. With everyone having a stake in the transparency processes coming into existence and successfully working, it might then be possible to require mandatory compliance for any holdout nations. The culmination of all of these efforts would be the achievement of a Cooperative Universal Transparency regime that would operate to ensure monitoring and verification of all nuclear weapons as well as inventories of SNM—over the whole world.

2.5. Developing New Monitoring Technologies to Support Expanded Demands

Cooperative monitoring has inherent limitations but offers unique benefits. Because inspections protocols are shared, the potential for deception and countermeasures is obvious. However, access rights afforded an international or multilateral inspectorate can extend well beyond those feasible by other means. As information gathering is done in the open, platforms that might be wholly impractical in a denied access context become feasible. Persistent surveillance opportunities are also expanded. The technical solutions pursued for cooperative monitoring must not infringe upon national technical capabilities, but the different design space and relaxed operational requirements can open up useful monitoring opportunities. Weapon design information must be protected and legitimate proprietary interests must be protected. Information barriers that allow reliable conclusions to be drawn by an inspecting party, while protecting design information and legitimate proprietary interests, must be developed and utilized.[18]

Research and development, ideally undertaken in partnership with other nuclear weapons and NNWS, along with the IAEA, must address several challenges, and in every case, take advantage of advances in information technologies. Priority should be placed on:

- **Information Barriers.** Information barriers that provide robust protection against unauthorized disclosures of sensitive information must be demonstrated. An ongoing level of effort is needed because vulnerabilities change over time as technology advances. More work is needed to ensure that the confidence required in a warhead verification measurement can be attained when such a barrier is employed.
  Authentication is essential to the process if trust is to be achieved. However, it has been best understood in a data transmission context, and is less well developed for a cooperative monitoring context. Approaches to functional testing, joint development, and random selection along with advanced measurement techniques should all be explored. A clear understanding of the measurement protocol to be utilized, along with the development of appropriately tailored hardware/software, is needed to identify and mitigate vulnerabilities.
- **Persistent Surveillance and Perimeter Monitoring.** Measures to provide enhanced and persistent surveillance within facilities, to monitor movements in and out of facilities, and on a limited regional basis will be needed in both arms control and international safeguards applications. Whether declared items and materials are of weapons origin, or derive from civilian activities, the ability to remotely monitor and understand nuclear activities in near-real time is needed. Maintaining continuity of knowledge can deter violations and provide a basis for more efficient verification.
  New methods for monitoring facility perimeters and penetrations are needed to verify facility designs and monitor relevant material movements. Transfers from facility to facility over a wide but known area must also be tracked. These capabilities can be applied on a routine basis or as part of enhanced monitoring requirements authorized under U.N. Security Council resolutions in response to cases of noncompliance, e.g., the gas centrifuge facility at Natanz. Such monitoring approaches could benefit from recent lessons learned in developing ISR technical and architectural solutions to support conventional warfighting.[19]
- **Real-time Process Monitoring.** Within a facility, efforts to improve material assay capabilities should be augmented by on-line instrumentation and process monitoring to provide the detailed information necessary to understand complex facility operations at enrichment and reprocessing plants. The large quantities of data generated through such approaches must be analyzed to draw relevant conclusions in a timely manner. This requires effective algorithms that correctly identify activities of concern without hampering legitimate activities.
- **In-Field Inspection Tools.** While accurate material measurements will remain important, IAEA is increasingly focused on investigations of noncompliance and assessing the nature of a state's overall nuclear program. Since the Persian Gulf War

and the revelations of the extent of the Iraqi nuclear program, the IAEA has shifted its attention to an examination of a state's actions as a whole. This has led to an increased need for investigatory tools that can be used for in-field measurements during an inspection, perhaps at previously undeclared locations. Providing as much information as possible to an inspector on past nuclear activities at a site, preferably while they still remain in the field, will grow in importance. Such forensic capability is also valuable in verifying cases of nuclear roll-back (e.g., Libya) and potentially the verification of dismantling or decommissioning in an arms control context. To assist investigation efforts in cases of suspected noncompliance, new tools are needed to support inspections in the field. Portable analytical capabilities that are capable of providing rapid and accurate sample analyses to inspectors should be further developed. Improved capabilities in this area help ensure that follow-up activities can be swiftly identified, limiting opportunities for concealment attempts.

2.6. Considerations for U.S. Nuclear Modernization Programs

As future arms control efforts will likely result in expanded access at U.S. facilities, weapons, and platforms, approaches for meeting verification objectives while limiting overall intrusiveness will be needed. An informal survey by the Task Force indicated that DOE/NNSA is aware of the concern and has introduced the issue into the programs for the new production facilities it is building. For example, a team from DOE's nonproliferation program and from the State Department had a favorable review with the designers of the new Uranium Processing Facility in the weapons program. The Task Force did not learn of any comparable considerations integral to DoD's programs.

2.7. Open Skies Treaty – Another Opportunity to Build Upon?

The Task Force also considered the merits of extant treaties against the future proliferation environment and found them in need of updating to apply them to a broader proliferation agenda before significant investments are made. An example that was considered in detail was the upgrading of the capabilities of the United States Open Skies Treaty aircraft. Based on the quality of the sensors allowed by the treaty, the Task Force would not recommend such a course of action at this time. The sensor specifications permitted by the treaty are outdated when compared with the need. In fact, the existing treaty requirements can be fulfilled by sensor information readily available from commercial imagery without the expense of flight missions or sensor upgrades. Therefore the costs of such an upgrade of the aircraft, which would be significant, are not justified at this time. The Task Force believes, however, that the original principles of the treaty remain valuable and could serve as a template for expansion to other bilateral agreements, but the compliance protocols should be updated before any new aircraft upgrades are considered—in spite of the fact that the Russians are upgrading their system.

2.8. Recommendations: Cooperative Regimes

International Cooperation and Transparency

State/AVC (diplomatic), DOE/NNSA/NA-20 (technical), and DoD/DTRA (operational) should develop a joint long-term international engagement plan aimed at the ultimate goal of international cooperation and transparency, and cooperatively implement their respective/shared responsibilities.

The Task Force recommends a 4-phase approach for expanded cooperation, each step of which would gradually evolve to "internationalize the transparency inspections." The Task Force approach could be enabled by expansion of the role of IAEA for assuming responsibility for the transparency responsibilities that ultimately emerge:

1. Bilateral, cooperative developments and evaluations among P-5 states, building on experience in historic allied partnership programs, as well as the CTR program with Russia;
2. Extension to all nuclear weapon states;
3. Expansion to nuclear materials transparency among major states with nuclear power generation;
4. Negotiation of a future Non-Proliferation Treaty (NPT "X") to bring in all nuclear weapon and material programs into a cooperative, multi-lateral regime.

The three lead offices cited should establish a Multi-Agency Roadmap in partnership with other relevant players in the USG. The effort should be spearheaded by a (U.S.) Mission Manager[20] to drive the realization and coordination for this initiative throughout the development process.

Key steps in support of the engagement plan and roadmap include the following:

- The activities outlined in the Road Map should be given priority within the participating Agencies, including clear definition of the anticipated step-by-step approach to realize the initiative.
- The State Department should begin to consult with foreign representatives, in order to prepare for their participation, starting with a rough schedule for joint efforts, and devise measures of effectiveness to ensure success through the distinct phases (P-5, nuclear weapon states, SNM possessors, etc.)
- Blue Team and Red Team efforts[21] should begin and be coordinated from the outset, to ensure that the proposals brought forward can meet the requirements to protect our own (and others) sensitive information from being revealed, while meeting the goals of accurate SNM characterizations.
- Major efforts should be mounted to adapt the proposed technologies into use, in the U.S. and in countries worldwide, to ensure that security measures are not compromised for either warheads or materials. (This will entail creativity and flexibility by both the inspected and inspecting parties to ensure that robust security information protection and material protection measures are retained.)
- Focused consultations with the U.S. Congress will be needed to ensure that all foreign interactions and participations meet U.S. Atomic Energy Act protections and

classification constraints, including establishing new Congressional "123" Agreements[22] for participating nations as required.
- As the transparency effort emerges, it will be critical to ensure (1) a high level of confidence in, (2) acceptance for use of, and (3) expressions of support for, the efforts by all U.S. departments and agencies responsible for U.S nuclear weapons and for SNM inventories in government and private custody. (The Task Force notes that advance buy-in and enthusiastic support for eventual achievement of a Universal Transparency Regime is crucial to prevent "pocket-vetoes" within the U.S. interagency forums, and later in negotiating the international agreements.)
- Finally, the Task Force recommends that a line item budget be sought for this important multi-year initiative, with *dedicated funding* for developing the key technology pieces, for creation of the operating systems that will be proposed for joint international use, and the cooperative experiments and trials with foreign nations.

Research and Development

DOE/NNSA/NA-20 should pursue an international R&D program in automated monitoring and reporting systems supported by information barriers and authentication to enable more effective and extensive materials monitoring. Trusted information barriers, capabilities for real-time process monitoring and in-field inspection and analysis capabilities should be developed.

DoD/Assistant Secretary of Defense for Nuclear, Chemical, and Biological Defense Programs ASD(NCB), DTRA, and DOE/NNSA/NA-10 should partner to develop cooperative options for asymmetric nuclear weapons security paradigms; e.g.,

- Sharable technical security principles, practices, and technologies for sites, materials, and components
- Site declaration and portal monitoring
- "Assurance Volume" concepts

U.S. Modernization Programs

In anticipation of treaties and agreements with more intrusive inspection regimes, State/AVC, DOE/NNSA, and DoD/AT&L should review current U.S. facility and weapon system modernization programs and instruct program managers, if necessary, to plan for accommodation of greater transparency measures. Methods for red teaming and performing vulnerability assessments should be developed and exercised routinely.

SECTION 3. UNILATERAL MEASURES: TRANSFORMING THE MONITORING FRAMEWORK

3.1. Introduction

Careful review of the ability to detect undesirable nuclear activities by non-nuclear states and non-state actors, as well as nuclear states experiencing internal instability or external threats, was a major part of the Task Force's effort. In general, prior reviews have come to two conclusions: that discovering such activities, if the perpetrator seeks to hide them, is a very difficult problem, and that the technical capabilities to do so are limited. This Task Force agrees with those conclusions, but believes there are some avenues of R&D worth pursuing to improve the current situation.

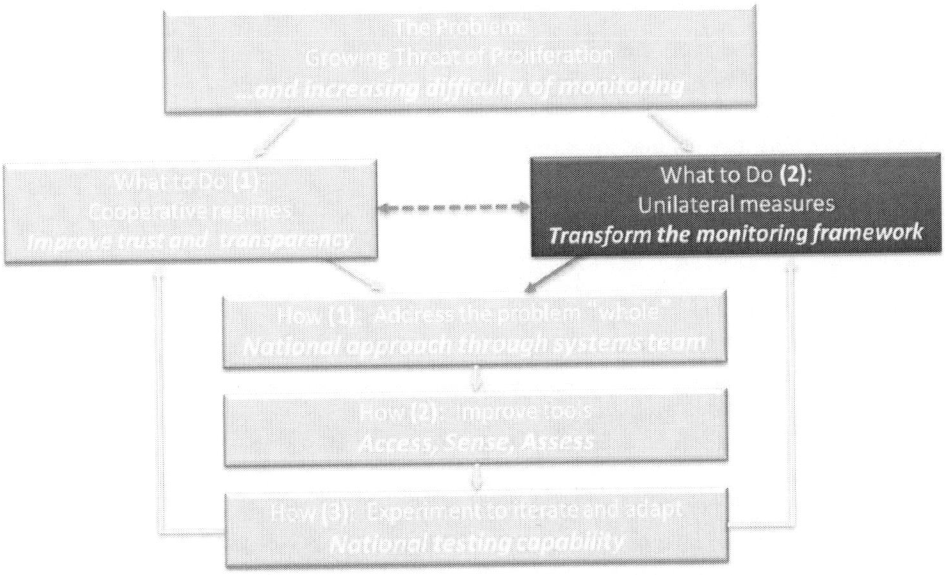

The national security community has faced such difficult problems before and generally found ways to achieve significant capability, if not entirely to solve them. Two examples are Anti-Submarine Warfare (ASW) and the recent problem of Countering Improvised Explosive Devices (Counter-IED). Both of these problems represented major threats and were difficult to solve, but a combination of technologies, some considered "out of the box," used in an end-to-end approach, was employed to achieve a satisfactory solution.

In its first phase during WWII, the ASW problem was significantly diminished by the advanced code breaking devices developed (the precursors of modern day computers), along with careful analysis of the patterns of operation by the German fleet. The second ASW problem, tracking the extremely quiet nuclear propelled and armed submarine developed during the Cold War, was addressed by the judicious blending of advanced large scale sonar arrays, discrete shadowing of threatening boats, and advanced algorithms to pull small signals from high clutter environments.

The IED problem has been addressed with a twofold attack: find the network and find the device. This has resulted in a significant mitigation of the threat. Both existing and new technologies have been brought to bear to address this problem. They can be categorized and characterized in the following six categories:

- **Persistence**: Holding the adversary at risk 24/7
- **Multi-INT integration**: One or more INTs to "find", one or more to "fix"
- **Advanced network analyses**: Tools to derive information from huge data sets

- **New cyber tools and techniques**: Cyber mining and geolocating to latitude and longitude
- **Improved SIGINT**: More signals, under more conditions

3.2. Characteristics of Monitoring Illicit Nuclear Activities

The Task Force agreed that *the guiding principle for monitoring to detect undesirable nuclear activity should be detection of activities as early in the planning and acquisition of a capability as possible* in order to provide the greatest number of options for slowing or reversing the effort. Difficulties arise from several factors—each one a challenge in itself, but in combination, as or more daunting than the ASW or Counter-IED problems. First, the number of actors and their geographic dispersion worldwide is large. Second, the observables are limited, typically ambiguous, and part of a high clutter environment of unrelated activities. Moreover, at low levels associated with small or nascent programs, key observables are easily masked and observation made more difficult without ready access by the full range of persistent intelligence systems. Third, and perhaps most significant, is that radiation phenomenology unique to SNM is not detectable at long standoff distances, further exacerbating the large area problem.

The signal-to-clutter characteristics are similar to those faced in ISR support to conventional warfighting and counterterrorism. The ISR architectural approach is based on a cueing principle that starts with general observations from multiple intelligence and surveillance sources and assigns ever more specific and precise assets to targets that appear to be of growing concern. Distinguishing characteristics of this approach are persistence, but not necessarily of the highest fidelity, widespread access to all the available data, allowing more "eyes on target" to detect anomalous behaviors or activities, and flexibility to rapidly reassign monitoring assets to "pop-up" areas of interest.

The Task Force believes there is promise in applying the conventional ISR paradigm to the proliferation monitoring problem. Unknown, however, is how well the ISR paradigm can be adapted to the spatial and temporal domains for monitoring suspect nuclear activities where long periods of observing normal patterns of life over large geographic areas will be punctuated by short duration anomalous events. Nonetheless, the impressive advances made in Iraqi Freedom and Enduring Freedom in support of locating both IEDs and the perpetrators warrant a serious effort to see how well these ISR technical and architectural approaches can be adapted to support proliferation monitoring.

3.3. A Cursory Examination of ISR Applied to Nuclear Monitoring

A closer examination of undesirable nuclear activities reveals two categories of focus for monitoring activities. Whether state or non-state actors, the two "headline" activities of concern are: 1) *obtaining* a nuclear device from existing stockpiles, or SNM from existing storage facilities; or 2) *developing* a nuclear device. In both of these cases, the primary objective should be to transition the initial monitoring problem (overwhelmingly large area and very small observable signature) into a more tractable one by either reducing the area that needs to be monitored or by establishing a more observable indicator of the activity.

The Task Force then dissected the problem into its simplest parts. The ways that a nuclear device could be *obtained* from an existing nuclear arsenal include:

- **Buying**: A large amount of money might induce the release of a few;
- **Stealing**: An insider could sneak one out of a facility;
- **Be given one**: Leaders of a weapon state could give one to a sympathetic or surrogate organization;
- **Recovering**: A lost device might be found;
- **Capturing**: A device may be taken from a storage facility or during transit.

The *development* of a nuclear device is in a relative sense conceptually simpler in that the developer must obtain or process SNM. Plutonium is more easily obtained, but weaponization more difficult. By contrast, HEU is more difficult to produce, but a device is relatively easy to fabricate.

With this simplified view of undesirable nuclear activities to be detected as early as possible, monitoring objectives can be stated as follows:

- Maintain an accurate accounting of existing nuclear material and devices in both declared and non-declared countries (see Section 2 for a phased path to achieve this objective);
- Detect and follow progress of suspected nuclear device development activity.

The Task Force then asked what the ASW and IED problems might teach us about this problem. The principal lesson is that detection using any single phenomenology or based on a single observable is not going to be successful. A multi-modal approach that goes after both primary and secondary observables is required.

Much of what was learned from the IED problem can be summarized as "find the network, find the device." Groups, not individual actors, carry out nuclear activities, much like the development and deployment of an IED. Coordination among several actors and organizations is required either to obtain a device or to develop one. Observing this coordination is the first indicator in the monitoring process and one for which many of the tools developed in dealing with IEDs should be well suited.

Exploiting the cyber domain should certainly be a big part of any nuclear monitoring effort. Both passive, depending on what is sent voluntarily, and active sources should be considered. Data gathered from the cyber domain establishes a rich and exploitable source for determining activities of individuals, groups and organizations needed to participate in either the procurement or development of a nuclear device. In fact, a set of new techniques, e.g. Advanced Graph Analysis (i.e., modeling relationships between data elements and data attributes as graphs to solve complex analytical problems), for exploiting these data have been developed and can be applied to the network of people possibly engaged in the activities associated with nuclear weapons.

Exploitation of the cyber domain to follow people, financial transactions, etc., is a critical first step in establishing a focus for technical monitoring based on other phenomenologies.[23] In other words, given the cyber cues, concentrated multi-INT collection and exploitation over a focused area is feasible. This is the basic process that has been used with success in recent areas of conflict where air and space collection platforms have generally enjoyed unfettered access. Moving forward, these same processes need to be applied in denied areas. Many of the new technology advances in data exfiltration, covert implantation, etc., hold promise for successful multi-INT collection and exploitation in non-permissive environments.

3.4. Recommendations: Unilateral Measures

The guiding principle for monitoring to detect undesirable nuclear activity should be detection of activities as early in the planning and acquisition of a capability as possible in order to provide the greatest number of options for slowing or reversing the effort. New ISR technologies, demonstrated in recent conflicts, offer significant promise for monitoring undesirable nuclear activity throughout the world. The nature of the monitoring challenge, however, is such that these technologies are most effective when applied in an integrated architecture. To accomplish that goal, the Task Force recommends that:

- The IC (led by the National Counterproliferation Center [NCPC]), DOE/NNSA, DoD/DTRA and Department of Homeland Security (DHS)/Domestic Nuclear Detection Office (DNDO), should develop a joint roadmap supported by the necessary systems analysis and engineering capabilities to implement an integrated, more comprehensive and responsive monitoring architecture for nuclear weapons activities worldwide, expanding upon the more general, but static approaches currently employed. As the legislated interagency lead for the Global Nuclear Detection Architecture (GNDA), DNDO should incorporate this more expansive monitoring architecture into the GNDA. The monitoring architecture should be structured to:
    - Create "corridors of observation" in multiple domains (geographic, commercial, individuals, financial…) based on open source and available multi-intelligence (INT) information;
    - Establish "patterns of life" within corridors that are suspicious;
    - Focus persistent monitoring assets on individuals and activities of greatest concern that emerge from patterns of life analyses;
    - Assemble and analyze data from all sources to support verification or identify previously unknown concerns;
    - Use results of analyses to iterate—i.e., to provide ever more focus for intelligence taskings, as well as provide guidance for needed improvements in technical and operational capabilities.

It should also identify lead and support roles and responsibilities for its implementation.

- National Security Staff (NSS) should request that DoD integrate the architectural elements into a global awareness system that:
    - Builds on lessons and experiences of successful national security capabilities, such as the Counter IED Operations Integration Center (COIC), NSA's counter-

terrorism capabilities, NCPC's counterproliferation efforts, and on DOE's Proliferation Risk Analysis Program;

  - Builds also upon the lessons and experience of the IC's Treaty Monitoring Manager[24], which at the height of arms control monitoring in the late '80s and early '90s proved an effective vehicle for coordinating the work of multiple organizations and collection activities;
  - Uses the Monitoring and Verification (M&V) national testing capabilities (described and recommended in Section 6) as the key operational assessment vehicle to test a range of scenarios.
- The IC should recast DOE/IN's historic "nth country" analytical capabilities into an "nth group" effort, adapted to the wider range of actors and designs accompanying both vertical and horizontal proliferation.
  - The "nth group" program should serve as a community asset for helping to characterize proliferation pathways from available data and advancing technology, and posit pathways to guide collection priorities.

SECTION 4. ADDRESS THE PROBLEM "WHOLE": A NATIONAL APPROACH THROUGH A SYSTEMS TEAM

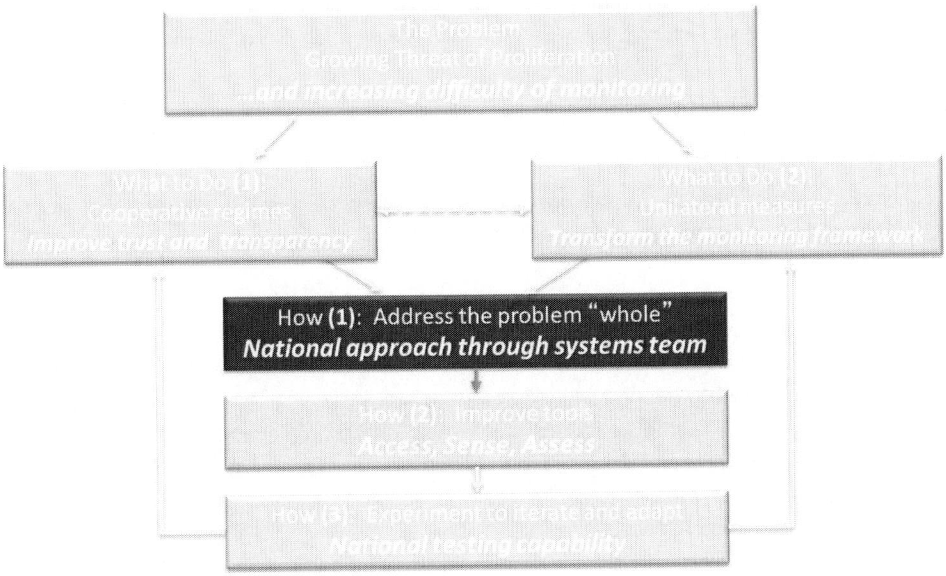

4.1. Introduction

Monitoring and verification for proliferation present a set of challenges that are difficult both technically and operationally. Progress and efficient use of resources in both domains should rely on solid, on-going analysis, but the analytical scope and complexity are no less challenging themselves. Many of the analytical challenges associated with M&V have been addressed in the past within the framework of historical treaties and agreements, which have often been unique to specific problem areas and definitions (such as detection limits, surveillance methods, and treaty details). As planners consider how the U.S. should address monitoring and verification missions in the future, new technologies, treaties, agreements, policies, and proliferators will change the landscape of capability requirements and continue to add complexity. The Task Force believes that experienced professionals in the area should not continue to address the problem as an extension of approaches used in the past treaties and agreements.

To better address the future monitoring challenge, the Task Force believes that there is a need to establish an enduring yet adaptable holistic methodology that encompasses the totality of the problem and solution space, along with the analytical capability for exercising that methodology, for guiding and facilitating the creation of roadmaps and plans for M&V capabilities as well as for evaluating developing capabilities. In other words, addressing the future M&V challenges demands a systems approach. Moreover, execution of a systems approach would be most effectively accomplished through a "national" team to help provide a high degree of unanimity and coherence of purpose among the numerous and disparate agencies involved.

The Task Force also recognized that while easy to recommend a systems approach, it is much more difficult to do—and do well. We therefore challenged ourselves to see if we could provide an example and potential starting point for the community to understand the problem in its full breadth, depth, and extent, and to offer guidance on the path forward. This Section provides an abridged version of those results, which are detailed more fully in Appendix A.

4.2. Elements of an M&V Analytical Methodology

Several key elements are proposed as necessary parts of an enduring M&V analytical methodology, as laid out in Figure 4-1.

- First, there is the "problem space," which consists of a set of frameworks that describe the M&V challenge *independent of any proposed solutions.* The problem space provides the basis that will serve as a common foundational understanding for solutions to be built upon and assessed for effectiveness.
- Second is the "solution space," in which proposed capability architectures are crafted. These architectures are collections of technologies, operations, and capabilities that work together to accomplish the goals and objectives identified to address the problem space.
- Third, there must be a Bridging Methodology which allows for the "back and forth" between the problem space and solution space; i.e., the tracking, integration, and trade-offs among objectives, requirements and architectural solutions to the problem.
- Finally, there must be a rigorous and repeatable portfolio decision methodology that allows for the values of the decision maker and his/her organization to be incorporated, and for defensible, tractable decisions to be made.

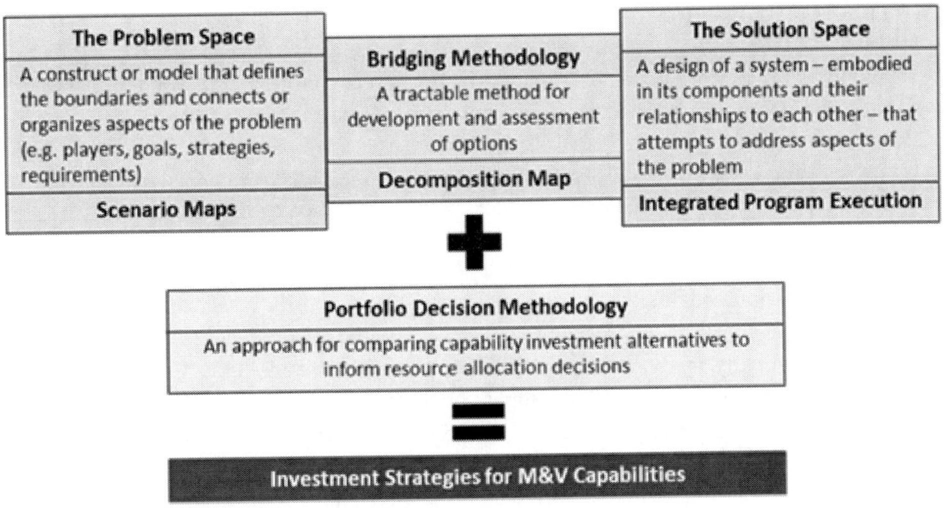

Figure 4-1. Elements of an M&V Analytical Methodology.

A common issue in the development of analytical frameworks has been the tendency to frame the problem and solution together. The conflation of "problem space" and "solution space" brings with it several consequences:

- It tends to promote metrics and assessments that support optimized subsystem or component performance vs. system performance by assuming they are one and the same.
- It tends to lead analysts to fixate on a narrower problem (such as detector performance) rather than assessing those issues in the context of the larger whole.
- It does not enable a common understanding of the problem space itself, allowing solutions to be framed in whatever manner the solution proponent deems suitable.

All of these symptoms run counter to enabling the decision maker to achieve his/her ultimate goal: rendering and defending a provably effective investment strategy for M&V capability development. Separating the "problem space" from the "solution space" and analyzing them independently can therefore yield benefits by eliminating or mitigating these issues.

To illustrate the point, consider that detection systems have often been placed at the forefront of the national strategy for detecting the illicit movement of nuclear threats in proliferation regimes. While undoubtedly detectors play a role in the solution, the bulk of current analytical activities are focused on detector system performance. As a result, the M&V problem of detecting illicit movements is often miscast implicitly as a detector problem. This can lead to the line of thinking that more detectors with better detector performance parameters must logically provide reduced risk – a statement that may or may not be true. More importantly, other options that do not hinge on detector deployments may in fact provide more cost effective mechanisms for risk reductions.

4.3. Proposed Problem Space Description

While there are several possible frameworks for describing the M&V problem space, the Task Force chose to use one based on scenarios, defined here as follows:

> A scenario is an evolution of the world through a series of incremental events from its current status towards an outcome of interest that is specified by the analyst.

However, the Task Force did not adopt a scenario-based planning approach because of: the pitfalls of drawing conclusions from a too narrowly defined set of scenarios, especially if they represent the analysts' or decision makers' "favorites"; or conversely, the unwieldy nature of systematically analyzing too many scenarios. Instead the Task Force developed a scenario framework that attempts to encompass a large portion of the potential scenario space in order to address the totality of M&V problem complexity. A scenario framework for scenario generation and analysis is desirable primarily because it enables the examination of a family of scenarios, rather than a small set of independent and specific scenarios. It also provides a systematic method for decomposing scenarios into discrete nodes and linkages, and capturing the interdependencies between individual scenarios. Finally, it lays the foundation for a bridging methodology, or systematic mapping between the problem space and solutions space, that enables increased traceability between planner objectives and solution performance.

The scenario framework developed by the Task Force for M&V started with the construction of a number of simple scenarios characterized by key nodes (events) and associated linkages. One such scenario, initiated by the collapse of a nuclear weapons state's security system, is illustrated in **Error! Reference source not found.**

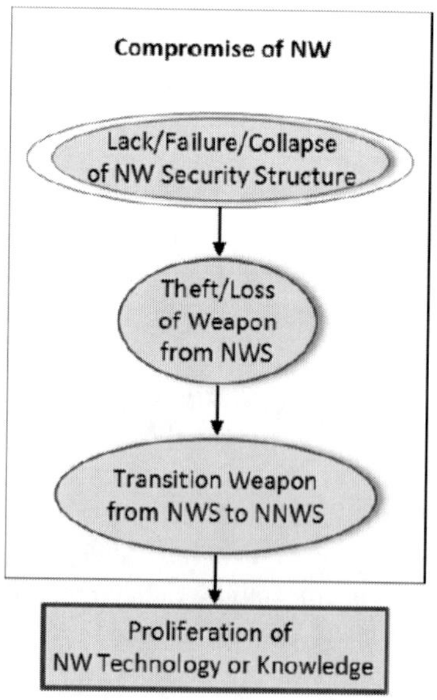

Figure 4-2. Example Node Sequence.

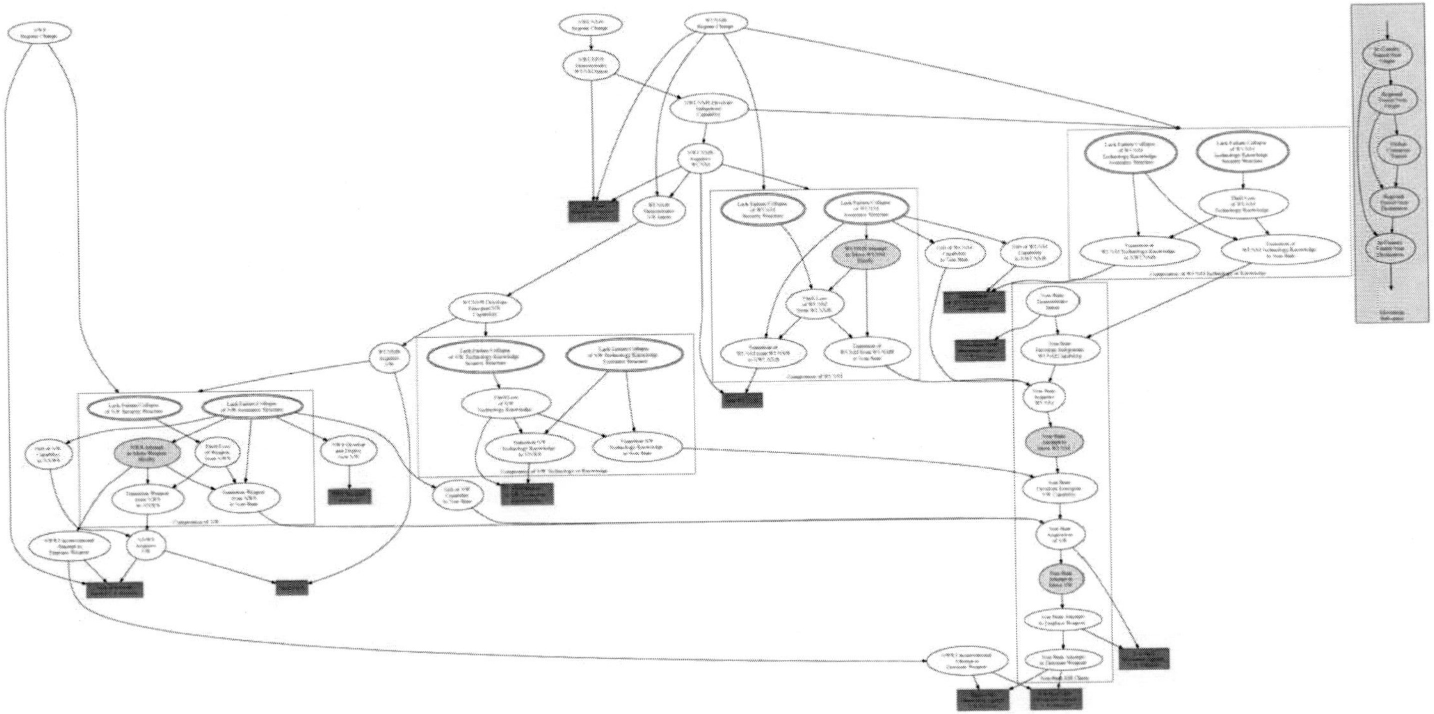

Figure 4-2. Scenario Framework of M&V.

The Task Force proceeded to construct as exhaustive of a set of such scenarios as it could think of, using this node-linkage approach.[25] The effort yielded both a practically complete description of the M&V challenge, as well as the identification of shared nodes and pathways among scenarios. The complete scenario framework (plus a readable, large format version in Appendix A) is illustrated in Figure 4-2.

Blue-colored nodes indicate "macro" nodes, or nodes that are easily decomposed into a common set of sub-nodes. The "movement sub-space" is examined as an example. This sub-space corresponds to movement through the origin region, across international boundaries, through global commons, and into the target region. The movement sub-space is laid out in the blue box on the right hand side of the framework diagram.

4.3.1. Analysis within the Scenario Framework

The scenario framework serves several roles. Most importantly, it exists as a common frame of reference for describing the M&V problem space. Narrower problem definitions, metrics, and objectives can be derived where appropriate through decomposition (discussed in Section 4.4.1). Scenarios for analysis can be generated by stringing sequences of nodes together. Any starting point and ending point can be selected, and a path through the network selected to connect them. From that string of nodes, a more complete narrative can be constructed.

The scenario framework can also allow for greater and more complete coverage in the design and analysis of solution architectures. A large family of scenarios can be analyzed by examining all nodes systematically node-by-node independent of end-to-end scenarios. An analyst can consider solution architectures that combat adversary success within a single node, and consider the collective impact it has on the complete scenario space by examining both upstream and downstream nodes. In addition, tradeoffs between architectures designed for different nodes can be compared in an end-to-end system performance sense. This will aid the assessment of the complete set of architecture components within a portfolio of defensive measures and allow for complex trades to be made. The scenario framework should be periodically reviewed and updated as appropriate based on real world experience and additional analytical studies.

While the scenario framework provides the structure for this kind of analysis, there is still the challenge of developing end-to-end metrics that are solution independent and common among all nodes. Further consideration of this issue is given in Section 4.5.1.

4.4. Bridging Methodologies

The method used to connect the problem and solution spaces is called a bridging methodology. It allows for a breakdown and prioritization of goals and objectives in the problem space into requirements and metrics for potential solution architectures. It also allows for the systematic aggregation of performance assessments and analyses into an overall picture of monitoring and verification architecture performance.

4.4.1. Proposed Decomposition Map Approach

The bridging method proposed by the Task Force is a decomposition approach that systematically maps problem space descriptions to prospective solution architect elements. The approach begins with the selection of any node in the scenario framework discussed in Section 4.3. The selected node is decomposed into sub-nodes required to add appropriate fidelity or resolution to the analysis. A system of decomposition layers is then constructed beneath the scenario nodes. Those layers, with increasing levels of specificity, are:

1. **Strategic Capability Areas** – This layer centers on core elements of the mission space associated with reducing risk.
2. **Functional Objectives** – Within each Strategic Capability Area, several functional objectives are articulated to capture high-level operational objectives that must be achieved.
3. **Tasks** – Each functional objective is further decomposed into a set of tasks. The tasks themselves are part of prospective solution architecture – i.e., tasks, just like objectives, are not universally defined, but proposed as part of a solution option.
4. **Assets** – Each task is accomplished through the employment of assets. Assets can include hardware, platforms, people, training, concepts of operations, and programs – essentially any capability that can be specifically invested in.

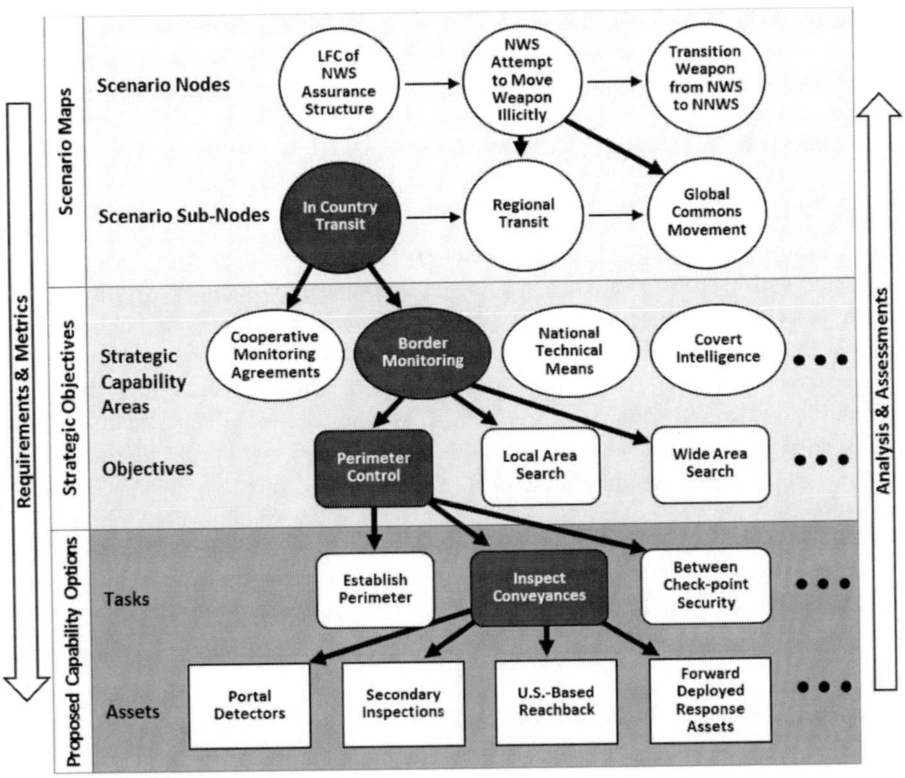

Figure 4-3. Example Decomposition Map.

Each node in the scenario framework proposed in Section 4.3 will have at least one unique decomposition map associated with it in a fully formed analytical effort. There are only a limited number of unique investible assets that may be incorporated in prospective solution architectures, and many assets are likely applicable to several different functional objectives and tasks. These observations imply that most assets will aggregate requirements from multiple scenario nodes, strategic capability areas, functional objectives, and tasks. Assets must be assessed against each set, and synergies may be identified and leveraged when designing solution architectures. Optimistically, the same asset may have sufficient performance and applicability across multiple tasks, functional objectives, strategic capability areas, and scenario nodes.

4.4.2. Decomposition Map Example

The overall decomposition approach described in Section 4.4.1 may be best described through the example illustrated in Figure 4-3. The example decomposition map begins with a focus on the scenario node "Non-State Attempt to Move Weapon." The scenario node is decomposed into more specific scenario sub-nodes; in this case the analyst is concerned with "In Country Movement" of a nuclear weapon or asset.

The next step of decomposition moves from the problem description (the top shaded area, labeled "Scenario Maps," in the decomposition map) into development of a solution architecture (the blue shaded area in the decomposition map). The scenario sub-nodes are decomposed into strategic capability areas dividing the problem space into discrete mission areas – in this case the strategic capability area of border monitoring is highlighted for further decomposition. Of several possible functional objectives for border monitoring, our example will consider perimeter control, which can be further decomposed into a set of tasks that must be accomplished in order fulfill that objective. These tasks and the assets associated with them represent a proposed solution architecture for accomplishing the objectives specified. For example, items that pass through that perimeter must be inspected to ensure that no nuclear contraband is passing across the perimeter.

As noted with the arrow on each side of the decomposition map, the requirements and metrics by which proposed assets are to be assessed have been derivatively defined from the previous layer. As assessments are made and analytical results obtained, they are aggregated into performance assessments of higher and higher layers. The derivation of a tractable set of metrics through this process is not a trivial task. Moving down the chain through strategic capability areas, functional objectives, tasks, and assets, metrics of increasing resolution and granularity are derived. At the asset level, the metrics center largely on performance specifications that are technology specific. These metrics are familiar in radiation detector assessments, for example, but on their own, are only implicitly related to the overall goals of risk-reduction. A metrics derivation process such as this places each metric in the context of the layer above it, explicitly linking it to overall architecture performance.

4.5. Portfolio Decision Methodologies

In order to render the analytical results that produce well-characterized architectural options from the approach described above into investment roadmaps, a decision framework must be established. While this section does not attempt to propose a decision framework, it does provide some considerations for doing so.

4.5.1. End-to-End Metrics

Risk (or risk minimization) is most often the implicit or explicit top-level metric for the decision maker. It provides a metric for endogenous trade-offs within the M&V problem space, allowing for the comparison of very different solution sets and examination of benefit between investments both within and across different components of the problem space itself. Utilizing risk as an end-to-end metric in the M&V problem space can also enable exogenous trades, as governments face economic challenges and must make tougher decisions about where to invest resources.

Utilization of risk as an end-to-end metric comes with a set of inherent challenges, however. Common criticisms of formal risk assessment methodologies in decision processes include:

1. **Conflating stochastic processes and adversary decisions** – Well characterized stochastic processes do not govern intelligent adversaries; instead, they make informed decisions. Although frequently used, probabilistic representations of adversary decisions are, for the most part, meaningless. However, characterization of uncertainty about adversary decisions in a probabilistic analysis can be beneficial, if carefully developed.
2. **Focusing on absolute values rather than relative impacts and sensitivities** – The absolute values of risk are, in most formulations, arbitrary, as they are built upon the assumptions and values of the analyst or decision maker for whom they are constructed. Additionally, the models upon which risk is calculated often cannot be truly validated.
3. **Inability to define "acceptable"** – A key component of making decisions in a risk-based framework is to define "acceptable" risks within the timeframe of the investment decision itself, something often difficult to achieve, especially when multiple equities are impacted.
4. **Examining risk trade-offs too narrowly** – In problems of high uncertainty, the error in risk calculations can be so large, that close trade-offs can be interpreted as essentially the same. Not having well-defined uncertainties can allow false comparisons to drive decision processes. At the same time, ignoring the uncertainty (or error bars) in the analysis denies they opportunity to prioritize potential efforts to improve understanding.
5. **Misuse of probability and statistics** – While it may seem like an elementary mistake, bad assumptions or interpretations, especially around dependence or independence, can lead to mathematical operations and inferences that may be numerically correct, yet meaningless—or worse, incorrectly calculated.

Given the difficulties that can make a robust risk definition in the M&V space difficult, it is possible to use measures that are proximate to risk when performing analyses. For example, other studies have proposed and formalized the replacement of probability in the classical formulation of risk with the use of assessments of the difficulty an adversary would face in executing a successful attack[26], when working with risk assessments of intelligent adversaries. Additionally, understanding the readiness level of an adversary and the expected time to attack can also serve as a useful risk proxy. Regardless of the exact risk metric or proxy used, risk (or some formulation of risk proxies) is a “necessary evil” as it allows for disparate approaches to be transacted together in cost-benefit trade-offs, and provides a consistent metric and method for analysis as new information and options become clear.

4.5.2. Balancing Risk

A key role of a portfolio decision methodology is to identify investments in capabilities that balance risks given the values of the decision maker. Investment risk includes not only cost and technical risk, but also institutional decision factors. Institutional decision factors are elements such as championship, mission, acceptance, and other factors that may prevent investments from being successful if not present. While not often considered in formal decision processes, they can have a large effect on the outcomes of decisions to acquire capabilities. These factors should be accounted for in any M&V portfolio decision methodology.

As an analyst considers capability options, a balanced portfolio may be very different given the values and risk tolerance of the decision maker and the investing agency. For example, R&D focused organizations may have a set of values that is more risk tolerant, and a balanced investment portfolio may be skewed towards the long-term, high risk, high payoff projects. At the same time, an operational component may seek primarily commercial-off-the-shelf solutions and a balanced investment portfolio may be skewed towards incremental, low risk improvements. Formal and quantitative methods for assessing the risk tolerance and the values of decision makers exist and should be extended in formulating an M&V portfolio decision methodology.

4.6. Proposed Analytical Capability

4.6.1. The White Team

The complexity of the M&V problem requires a different analytical approach, one example of which is the scenario framework described above. In turn, the complexity of that analytical approach points to the need for a focused and sustained effort to develop and exercise the framework, the bridging methodology, requirements and metrics. The recommended approach is an analytically focused team sponsored by the appropriate government agency(ies), and populated with technical and policy experts. This team is neither a “red team” nor a “blue team,” but rather an independent and unbiased body, focused on a high-level and comprehensive analytical effort to tee up options for decision makers with a clear eyed assessment of the pros and cons of each option. As such, the team should exist independently of operational or acquisition organizations. It could and should, however, utilize findings, analysis, and data from both red and blue teams. This team, which the Task Force labeled the “White Team,” would have a charter that includes:

- Fully developing and stewarding the problem framework, a bridging methodology, the associated strategic requirements, and metrics;

- Participating in test-bed activities, providing data requirements to further analyses, and specifying test cases;
- Presenting policy and acquisition options to implementing agencies for decisions;
- Working with trusted international partners to examine and exchange best of breed strategies and architectures.

As steward for the problem framework and the bridging function, the white team must integrate from the comprehensive understanding of the problem space to the pool of candidate solutions. The white team should accomplish this through not only its own intellectual endeavors, but also by working closely with a "Solution Provider Team," comprised of laboratories, contractors, research institutions, and academia, who together can provide a rich set of candidate solutions. These are the organizations that should be at the forefront of technologies, systems, and their integration into operations. The solutions provider team has the charter of:

- Proposing technical and non-technical solutions in response to strategic objectives and requirements, as well as the focus for R&D when existing solutions are inadequate;
- Preparing performance analyses to serve as first pass component specific studies to feed the White Team (the "ho-ho" test);
- Working with trusted international partners to develop best of breed solutions.

While the individual members of the solution provider team may have real or perceived biases or conflicts of interest, the overall solution provider team will have a more balanced perspective, and any recommendation from them must pass through the independent filter of the White Team.

4.6.2. Implementing the White Team

Two factors are required for implementing the White Team: 1) the identification of a cadre of analysts who see a strong opportunity for a valued career experience, if not profession; and 2) the establishment of an institutional mechanism that can ensure the team's access and impact across the interagency.

Regarding the first factor, the Task Force discovered that the experienced systems analysis workforce of the Cold War has largely disappeared, just as it has in other areas related to nuclear weapons. Moreover, although there have been significant successes in nonproliferation and counterproliferation analysis, these approaches have tended to focus on problems bounded by treaties and agreements, or by specific actors of interest. They do not easily scale to meet emerging challenges across the globe. The Task Force has attempted to portray how much more complex the environment is today and into the foreseeable future. As a result, new approaches and new tools are called for. The leading agencies therefore should understand that part of their charters should be the re-growth of knowledgeable professionals, especially systems analysts, to support threat assessment, trade-off studies and experiments, and investment prioritization. The second factor for a successful White Team is the institutional home and support that it will require in order to maintain both investment and operational functions throughout the interagency. The Task Force debated a number of mechanisms:

- Establishing a "czar"
- Appointing an executive agent
- Instituting a "holding company" model, similar to the On-Site Inspection Agency
- Assigning a mission manager (whose role is discussed in the next Section)
- Identifying a coalition of the willing, coordinated through an Interagency Coordinating Committee
- Utilizing the "Ungroup" model, which worked so effectively as an ad hoc, high level interagency group through the negotiation and implementation of the original START treaty and other arms control agreements.

Each option has its pros and cons within the context of the monitoring and verification problem, and none seemed well suited or sufficiently innovative to address the problem as the Task Force sees it; something more unique would be needed, at least initially, in order to protect the White Team while it establishes itself. The Task Force modeled its idea after the Phase One Engineering Team (POET), which was established within a couple of years after the standup of the Strategic Defense Initiative Organization (SDIO) at DoD in the early 1980s. The POET was populated by successive generations of some of the top talent from the Federally Funded Research and Development Centers (FFRDCs) and national laboratories; individuals were independent of specific acquisition programs. The POET reported directly to the SDIO director and had sufficient influence to establish the baseline threat assessment and balanced architecture options, as well as more specific assessments of proposed technical capabilities. Adapting the POET model into an interagency environment will require some variant of either a lead agency with accountability to partnering agencies (e.g., a "board of directors"), or governance by a special interagency board, chaired by the appropriate lead in the NSS or selected on a rotating basis among the participating agencies. The Task Force realizes that this approach is highly unusual, but also believes that other options fall short given the complexity of the problem, the need to think through new and innovative solutions, and the number of agencies and interfaces that are needed to make progress.

4.7. Recommendations: Addressing the Problem Whole

As the lead agencies, State/ASD/AVC, DoD/ASD/NCB, DOE/NNSA and IC/NCPC should create the processes and oversee the following steps:

- Establish the "White Team" whose charter is to characterize the comprehensive Monitoring and Verification Framework, relating threat events and actions to monitoring requirements, both cooperative and unilateral.
    - The White Team should be assigned to a "home" agency, but supported and governed by an interagency "board" whose members have sufficient authority to influence any needed changes in strategy and program directions at their home agencies.

- Ensure a common understanding among agency leads for addressing all aspects of the framework, including policy, diplomacy, operations, and RDT&E.
    - It is critical that those agencies involved in implementing M&V capabilities maintain a high degree of unanimity on how M&V problems and challenges are characterized, and how they will be addressed in order for the White Team to be both effective and sustainable. The adoption of common frameworks (such as that proposed in this Section and report) can contribute to developing a common understanding of national strategy, goals, and pathways to accomplishing those goals.

The foundation established with the two steps above leads to the following implementation steps, again under the charge of the leading agencies:

- Adapt or create integrated implementation plans. No single agency has purview over the totality of the responsibilities in the M&V mission, but integration "across the seams" is required.
    - The NSS, with support from OSTP, should assess progress annually.
- Establish an institutionalized interagency planning process that evolves with the threat.
    - The nature of the M&V problem and solution spaces has changed considerably over the past decades. Undoubtedly, the challenges and proposed solutions will continue to evolve in the coming years. Therefore, not only must the analytical framework used by the White Team evolve, but the plans and priorities across the government must evolve as well.
- Enlist a wide range of contributors.
    - Historic contributors such as the DOE national laboratories remain critical to addressing the nonproliferation problem. However, the Task Force believes that the agencies should enlist a wider array of performers (other labs, contractors, academia) to address the full complexity of this problem space that will call for new, as well as improved, tools.

Section 5. Improve the Tools: Access, Sense, Assess

5.1. Introduction

The second "how" in the Task Force's assessment, illustrated in Figure 5-1, addresses the "tool box" to support proliferation monitoring and verification. Figure 5-1 conveys the three principal areas for effective monitoring, each of which is in need of improvement:

- **Access**—global in a range of possible environments;
- **Sensing**—through radiation detection, post-event collection and analysis, and application of new modalities;
- **Assessment**—involving exploitation of massive and disparate data sources.

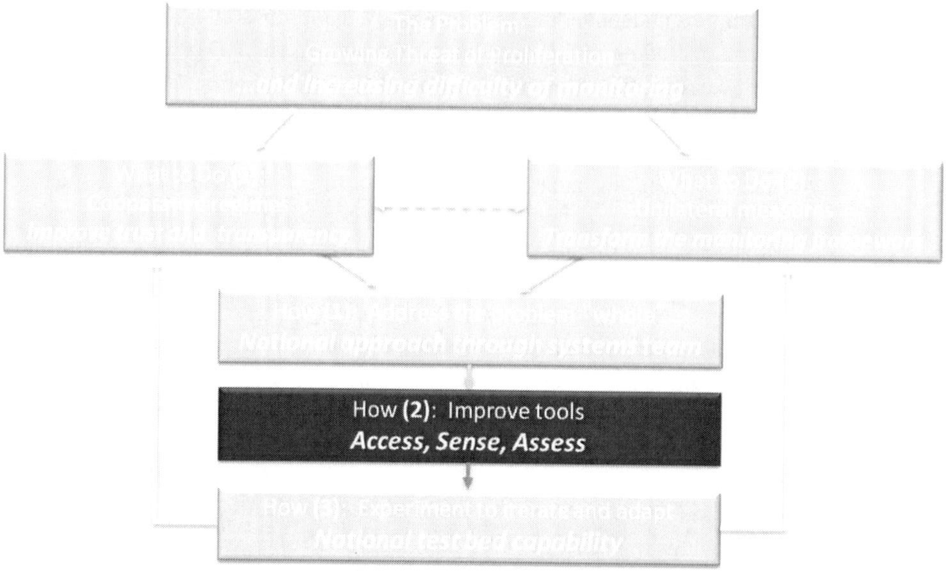

Further, as illustrated in Figure 5-1 and discussed earlier in Section 1, integration and iteration of monitoring results, within and across the full scope of monitoring environments, will be required if monitoring is to support verification.

5.2. Access Globally

Improving access is essential for making progress against the low and/or obscured signal of incipient proliferators. A principal benefit of expanded cooperative regimes, as discussed in Section 2, is the access it affords. However, many scenarios could require access in areas where access is difficult. Improvements across the board are needed.

5.3. Sensing: Radiation Detection

The detection of radiation emitted by enriched uranium or plutonium coupled with the identification of the source of the radiation has been regarded as the gold standard for monitoring SNM in process, in transit, and in storage; as material to be manufactured, as components for, or integrated in reactors or weapons, as spent or retired material. The radiation usually comes in the form of gamma rays and thermal or fast neutrons. There are a number of different detector systems in use and many more under active research and development. This section of the report describes the range of technologies and applications being pursued, as well as the role of the several government agencies that develop or use radiation detectors.

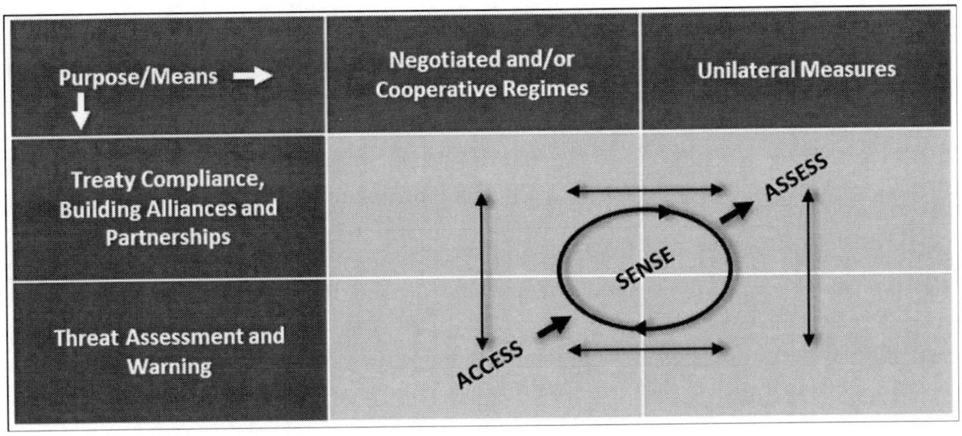

Figure 5-1. Improve Tools: Both Cooperative and Unilateral Measures Rely on Monitoring Tools to "Access-Sense-Assess-Iterate".

Material	Shielding	Distance (m)*	Distance (m)*
		Neutrons	*Gammas*
12kg HEU	3cm Tungsten	0.2	<0.4
4kg Pu (weapons grade)	3cm Tungsten	25	<0.6

* The distance quoted is that at which the passive signal equals the nominal background rate for handheld/backpack detectors.

Figure 5-2. Physics of SNM Detection.

5.3.1. Applications and Limitations

There is a wide range of potential applications for radiation detectors in the general area of monitoring and verification. Such detectors play an integral role in treaty verification. They are an essential element of on-site inspections for treaties and IAEA–type inspections. They are used to confirm nuclear material types and inventories under bilateral or multilateral agreements below the formal level of treaties. They are required for scanning at border crossings and ports, and are the confirming step in any search for suspected hidden SNM or weapons.

The detector technology of choice is dictated by several factors. Most importantly, the laws of physics and the limits of engineering art set bounds on what can be detected in various operational scenarios (see Figure 5-2). Detection and identification are relatively straightforward when access to potential sources of SNM is relatively unconstrained by distance or time. However, this is usually not the case. More often the detector system, including data processing and display, needs to be portable, relatively low cost, and easily operated in stressing environments with limited access. In every important application there is a trade-off that must be made among all of the possible characteristics that would be possessed by the ideal detector.

5.3.2. Detection of Special Nuclear Materials (SNM)

Detection of SNM can be accomplished passively by measuring natural radiation from the isotope in question, or actively by bombarding the isotope with radiation probes and measuring the resulting emissions. Passive detection is often limited by the low intensities of naturally occurring radiation from the source, the interference produced by background radiation, and the obscuring effects of shielding (including self-shielding in the object). Active detection usually has a much higher signal to noise ratio but requires complex interrogation equipment (frequently large and expensive) and often involves radiation safety issues. In addition to the passive or active question, detector design entails choices on a wide range of characteristics: cost, size, portability, radiation probe (for active detection), sensitivity, efficiency, energy resolution, discrimination, directional capability, imaging features, and features tied to the operational scenario of interest.

5.3.3. Status of R&D for Various Applications

The science and practice of radiation detection is many decades old and the most commonly used detector materials have changed little for much of that period. The field was somewhat rejuvenated after the end of the Cold War with the potential for new treaties as well as the goal of accounting for SNM throughout the world. The events of September 11 accelerated the need for new detectors and in particular, broadened the need for technologies that could detect hidden SNM in a variety of environments. Many of those new approaches and instruments are still in the development stage, but it is likely that a modernized set of detectors will be deployed during the coming decade. In all cases, these represent quantitative but evolutionary improvements in capability rather than revolutionary new techniques that overcome longstanding range limitations. Nonetheless they will greatly enhance the ability to monitor ports, border crossings, and vehicle cargoes for hidden SNM. What remains elusive is practical detection schemes at ranges greater than 100m and identification of illicit activities involving SNM.

There are copious reports summarizing the state of the art and future research directions in radiation detection.[27,28,29] Instead of repeating and summarizing the extensive discussions found in those documents, the Task Force chose to summarize the status of radiation detection technology for the main applications of interest and refer the reader to those documents for much of the background technical information.

Close-in Monitoring of SNM

Applications include on-site and IAEA-type inspections where close-in access is possible. In this case, there is a need for improved measurement systems for identification and quantification of SNM in a variety of material types and configurations. Examples of promising research areas include high-resolution gamma-ray spectrometers operating at room temperature; improved neutron coincidence counting and neutron multiplicity measurements; correlated measurements of fission gammas and neutrons; and advanced micro-calorimetry.

Hidden SNM at Ports, Borders, and in Vehicle Cargoes

As noted previously, a tremendous effort to develop radiation detection equipment for these applications has occurred since September 11. Passive gamma and neutron instrumentation has been deployed overseas (e.g., as part of DOE's Megaports program). The Department of Homeland Security has deployed fixed-site radiation detection and radiography equipment at U.S. ports and borders. Portable radiation search equipment has been shared with partners through the Proliferation Security Initiative. The next generation of radiation detection instruments with improved performance for these applications is in the pipeline of testing and qualification for deployment over the next several years.

Operations in Contested Areas

A variety of passive gamma and neutron radiation detectors are now available for operations in denied areas, primarily for troop protection where the presence of radiation is anticipated. Portable radiation detection equipment for aircraft and ground vehicles has been demonstrated under battlefield conditions. The concept of operations for these systems relies on external cueing that radiation sources could be expected in the immediate vicinity of operations. The next generation of radiation detection systems for this application may include large-array gamma and neutron detectors, passive neutron and gamma ray spectral imagers, and possibly active neutron and photon (bremsstrahlung) interrogation sources. If successfully developed and deployed—and the Task Force cautions that much work will be needed to do so—such active systems could extend the useable detection range out to ~100 meters depending on the quantities of SNM and associated shielding of the objects in question.

Finding Loose Nukes or Identifying Theater Nuclear Weapons

These continue to be the most challenging problems for detection and identification of SNM sources. As noted elsewhere in this report, radiation detection by itself will play a limited part in the solution to these challenges. A networked search and discovery architecture that includes non-radiation sensors along with near-real-time data and information fusion and dissemination will need to be developed. Radiation sensors would be engaged only during the end game when a likely location for a hidden weapon or SNM has been identified. At that point, radiation sensors would be used to identify and characterize the source before interdiction by personnel or destruction by kinetic munitions.

Longer Range Detection

Active interrogation for stand-off detection using high energy photons to stimulate fission has demonstrated detection of delayed neutrons. Interrogation distances up to 100m have been demonstrated, but detection distances have been limited to tens of meters. Long range (100m to kilometers) stand-off radiation detection and identification of SNM must await future breakthroughs. There is little chance of a deployed technical solution in the next 10 years.

5.3.4. Agency Responsibilities and Inter-Agency Coordination

The agencies with principal responsibilities for technical development as well as operational deployment of radiation detection systems are the Departments of Defense, Energy, and Homeland Security. The DoD must protect its bases and other sites, be prepared for detecting radiation sources in hostile environments, carry out on-site inspections and (ideally) detect SNM from standoff distances (> 100 m). The DOE has a myriad of different programs that involve radiation detection including non-proliferation, monitoring of material in the nuclear fuel cycle, operations at international borders and ports, and searches for suspected loose material or weapons. The DHS has the lead role in U.S.-based operations including borders and ports and for activities in public U.S. environments. There is a four-part MOU among those three agencies plus the DNI for coordination of nuclear detection R&D, and an interagency working group (IWG) led by the President's OSTP. These mechanisms help provide substantial information exchange and communication among the workers in the field and has been successful in avoiding unnecessary duplication of effort, but the diversity of technical and operational needs makes it difficult to set (or infer) overall research and development priorities. Novel concepts and new materials are continually being proposed and should be evaluated within each of the major government programs because it is possible that one may produce breakthroughs in cost and/or performance, but the Task Force found none that were sufficiently mature to warrant a major ramp-up in investments.

The major shortcomings across the federal efforts are two: (1) there is no integrated mechanism for setting R&D priorities across the programs or for allocating funding along such lines; and (2) operational requirements to measure the readiness of a given technology or operational system for transition to a specific application are often lacking, such that much of the radiation detection technology that is developed is never transferred to operational use. The task force believes that a more unified "national" technology roadmap can aid both the individual agencies and the efficiency of the overall federal programs. In addition, understanding the interplay between possible radiation detection capabilities and the broad range of possible (other) monitoring technologies, operations, and scenarios should be improved within both the R&D and operational communities.

5.3.5. Key Findings: Radiation Detection

The task force did not try to down-select or set research and development priorities among the various technologies since the wide range of applications often calls for differing solutions. Rather, it focused on identifying mechanisms that would enable the various agencies to do their work more effectively and also produce a better overall federal approach to radiation detection. Our specific findings are as follows:

- Operational requirements to measure the practicality and readiness of a given technology or operational system for transition to a specific application are too often lacking. As a result, much of the radiation detection technology that is developed is never transferred to operational use.
- Technology and concept of operations (CONOPs) are both workable problems for New START treaty verification. Intrusiveness and information protection must be addressed.
- Beyond New Start, the requirement for monitoring will begin to emphasize individual warhead or bomb counting; solutions that simultaneously verify the presence of a warhead without revealing sensitive classified design information will be needed.
- There are technically plausible approaches at ports and borders for detecting plutonium and radiological materials; shielded uranium requires continuing research. Several systems are currently in the pipeline for deployment during the next several years. All methods involve trades among CONOPs and cost-benefit choices, which must be resolved before deployment.

- Radiation detection will have a limited role in finding loose SNM or theater nuclear weapons; however, when a suspect object is found, radiation detectors should be able to identify and characterize it.
- Managed access at national nuclear facilities and test ranges is needed for technical and operational development of radiation detectors.
- There is little chance of a deployed technical solution for longer range (>100 m) detection of SNM in the next 10 years.

5.4. Sensing: Post Event

The Task Force was originally asked to look into capabilities for nuclear forensics but decided not to pursue the topic, with one exception, because of a focused National Academies committee which was completing its work during the course of this study.[30] The one exception was an assessment of the USAEDS, operated by the Air Force since the late 1940s. The full system, comprised of a global satellite surveillance constellation and a seismic-hydroacoustic network, coupled to the International (seismic) Monitoring System, also includes an airborne sampling system that is vectored to the area of a suspected explosion to collect radioactive debris.[31] The samples are returned to ground laboratories for detailed radiochemical analysis. The analysis can take days to weeks.

5.5. Sensing: Additional Modalities

Section 3 introduced the need for a significantly expanded monitoring architecture to address the challenges of both horizontal and vertical proliferation. The task force highlights here some of the particular sensing capabilities that should be assessed and/or integrated into the overall architecture.

5.5.1. The "INTs"

Many of the various intelligence collection capabilities (HUMINT, IMINT, MASINT, OSINT, SIGINT, etc.) have historically played a strong role in the detection and assessment of larger nuclear programs. Applications outside the nuclear proliferation domain in which signal-to-clutter is extremely small have been significantly improved, especially from an exploitation standpoint, during the last decade of warfighting. Crossing back into the nuclear realm and focusing on small, easily hidden or nascent programs will be important to meet the monitoring challenges identified in this report.

5.5.2. Scientific and Commercial Imagery

A major scientific and economic trend exploitable for increasing the capacity and persistence of monitoring coverage worldwide is the spread of satellite sensing capability throughout the world and into the commercial sector. Governments and nongovernmental organizations in over 50 countries now operate nearly 1,000 active satellites in earth orbit. Two hundred of those satellites are engaged in earth observation. This collection of satellites has several attributes of potential utility to monitoring:

- Observation of earth from satellites is no longer just within the purview of compartmented defense and intelligence programs. Information is collected by a growing number of states and companies and shared widely for academic and commercial purposes. This makes information from space, which is relevant to arms control monitoring and proliferation concerns, more easily shareable and accepted in international fora.
- International monitoring bodies themselves are increasingly engaged in the use of imagery from space. The IAEA, for example, has established a Satellite Imagery Analysis Unit. IAEA presentations note that commercial satellite imagery is now routinely used as an integral part of the safeguards system by which IAEA seeks to monitor independently the correctness and completeness of the declarations made by states about their nuclear materials and activities.
- There is a growing body of academic and non-governmental organization (NGO) literature that uses imagery from commercial or declassified sources to analyze activities or installations of proliferation concern. This literature identifies methodologies for the employment of information from satellite systems to improve monitoring regimes and has provided crowd-sourcing techniques for the detection of activities and places of monitoring concern. These methodologies employ readily available visualization tools and historical imagery available from websites such as Google Earth. Some observers, with some hyperbole, have referred to these techniques as "a new era in global transparency."
- Despite the funding vagaries of the commercial satellite industry, the number of commercial remote sensing platforms seems likely to grow. This growth will be accelerated by increases in hosted payloads, in cube- and mini-sats, and in new business enterprises offering innovative lift for payloads, including suborbital spaceplanes.

This proliferation of commercial remote sensing satellites does have some limitations and drawbacks:

- The increase in the volume of data and the reports of nongovernmental entities analyzing commercial imagery may introduce additional noise into U.S. and international monitoring systems. Some experts are concerned that bad data and bad analysis could increasingly tarnish or mask more reliable data. Moreover, these experts are concerned that such data and analysis would divert government professionals from more important tasks in the quest to corroborate or refute commercially derived claims.
- The greater availability and dissemination of data will constrain the ability of the United States to keep its options open in dealing with potential violations of international agreements or other threatening behavior. The distinction between the gathering of information through monitoring and the making of political judgments in the verification process will erode, and raw information gathered by commercial or international sources may become matters of public discourse that could hamper quiet resolution of noncompliant or ambiguous activities more difficult.

  On balance, however, it is the judgment from the task force that more information from remote sensing systems, both commercial and dedicated national assets, is better than less information. The increase in information from commercial systems adds to the importance of recommendations, made in Section 3, to improve the United States ability to process and analyze information proactively and to find weak signals of threat or noncompliance even within the most cluttered and noisy environments. The ready availability of commercial imagery and other open source data also increases the ability of the United States to share or discuss data with international partners in ways that serve to protect the most perishable of intelligence sources and methods.

5.5.3. Open Source Center (OSC)

The OSC provides accessible reporting on foreign political, military, economic, and technical developments in over 80 languages and from more than 160 countries. The OSC, largely based on the former Foreign Broadcast information Service, uses open sources such as radio, television, newspapers, news agencies, databases and the World Wide Web. Within the OSC is the Counterproliferation Program (CPP) office charged to track priority CP issues through various public media. Since the establishment of the CPP in 2007, the OSC has provided, by many accounts, excellent support with limited resources to the IC's counterproliferation mission.

Open Source (OSINT)/Social Media/Networking is the area that is undergoing perhaps the most revolutionary change within the IC and across the globe. "The poor man's intelligence community" is now available to anyone with access to the Internet. The continual launch of commercial space-based sensors of every phenomenology (e.g., electro-optical, multi-spectral imaging, hyperspectral imaging, synthetic aperture radar, and eventually full motion video) has meant an explosion in source data originating from outside the national security community – which is often difficult to process, exploit and disseminate. Nonetheless, these and other geospatial (or geospatially-enabled) data are becoming increasingly available around the world, and are helping to shape and reshape GEOINT and Open Source as new modes of high-value "intelligence." Cell phones, Smartphones, tablets and other like devices are increasingly able to sense location, giving rise to a new class of applications. In this class of applications, the line between Open Source and GEOINT will become more blurred. Eventually, as all open source information becomes more anchored in place and time, one will see Open Source transform how GEOINT is prosecuted and used.

5.6. Assessment: Data Exploitation

The Task Force focused on two emerging enablers for exploiting the increasingly vast amounts of data available at the heart of the assessment function: big data analytics and crowd sourcing.

5.6.1. Big Data Analytics

A popular term for an area that is growing rapidly in R&D investments, big data refers to collections of data sets so large and complex that until recently they were impossible to exploit with available tools. One can successfully argue that the era of big data arrived quite a while ago for the intelligence and military communities with the collection of data from sensor and other sources of information that overwhelmed the capacity to exploit all, or even most, of it. Technologies are rapidly evolving commercially to enable the storage, access, computational processing, etc., of such large data sets, but will require different "data centric" architectures and more importantly, new analytical tools for exploitation. A major tradeoff— or limitation— is emerging in big data analytics with the realization that the transmission latency among storage devices in a cloud based architecture is overwhelming the processing time for computational operations. As such, the analytics need to stay near the data; i.e., in a "back to the future" sense, large banks of co-located processors will form the big data processing architecture of the near future in those cases where the data is growing and/or changing rapidly. The strategic monitoring capability argued for in this report is one such example.

5.6.2. Crowdsourcing of Commercial Imagery

The use of crowdsourcing for nuclear monitoring offers the potential of increasing analytical capabilities without necessarily a substantial increase in the funding for specially designated offices for counterproliferation analysis. For example, the Task Force believes that crowd-sourcing of commercial imagery is one avenue that should be pursued. Imagery analysts focusing on counterproliferation at the National Geospatial-Intelligence Agency (NGA) and other organizations could focus their efforts on the higher priority and time-sensitive CP requirements, while crowdsourcing could be implemented for less critical imagery analysis requirements. The viability of the concept has been tested in several recent examples.[32,33,34]

The use of crowdsourcing of commercial imagery analysis does come with one major concern identified above in the discussion of the use of commercial and scientific imagery, namely the quality of both the data and analysis. A vital part of any commercial imagery crowdsourcing process has to be a thorough "quality assurance" process. There have already been major analytical errors made by untrained imagery analysts who have published openly.

A recent article in the Washington Post about work done by Georgetown University students analyzing the network of tunnels in China to hide their missile and nuclear arsenal is a good example of both the pros and cons of crowdsourcing.[35] Using open source information for the past several years—the Internet, local Chinese news reports, Google Earth and online photos posted by Chinese citizens—the students have published a far-reaching paper that challenges assumptions made by the IC on China's nuclear weapons capability. However, there is an extensive debate on the accuracy of this report, compiled by untrained intelligence analysts without access to classified data, which could have serious political and military implications. Many nonproliferation experts question the veracity of the report, citing how a semi-fictionalized Chinese TV series is used as one of the intelligence sources.

Whether the report is completely accurate or not, this event provides a "proof-of-concept" on how crowd sourcing can be used to augment limited analytical capacity. The IC should establish a process which codifies crowd sourcing as an additional area for research related to nuclear treaty monitoring issues, and ensure that a non-prejudicial process is established whereby open source and mainstream intelligence assessments can be reconciled.

5.6.3. Iteration: Proliferation Monitoring Management

Figure 5-1 illustrates the importance of integration and iteration in the access-sense-assess cycle. The Task Force notes that this responsibility should fall to the DNI, who bears a special responsibility for pursuing and integrating sensitive intelligence sources and methods, open-source information, and data provided through arms control-related information exchanges and inspections. In carrying out this role, the DNI can draw upon the lessons and experiences of the Intelligence Community's Treaty Monitoring Manager.

At the height of arms control monitoring in the late 1980s and early 1990s, the position of Treaty Monitoring Manager (TMM) was an effective vehicle for coordinating the work of multiple organizations and collection activities. As noted earlier in this report, the number of actors of both horizontal and vertical proliferation concern, and the increased geographic scope of activities of concern, have become too large to understand and anticipate developments of potential threat to the United States within treaty sanctioned regimes alone. The Task Force believes the TMM model could be effectively adapted to a Proliferation Monitoring Manager position with responsibilities to proactively orchestrate the collection, fusion, analysis, and dissemination of information vital to understanding the overall horizontal and vertical proliferation threats to the United States and its allies, as well as providing information relevant to verification judgments about the compliance of state parties to international agreements. This position, working with other National Intelligence Issue Managers, should explicitly act to create synergies among collection modalities and information means and to disaggregate tough overall monitoring problems into manageable problem sets. The Proliferation Monitoring Manager, as the Treaty Monitoring Manager has historically done, would play the leading role in the U.S. Government for orchestrating the timing and focus of the most sensitive intelligence collection with negotiated and overt activities such as inspections, data exchanges, cooperative measures, and diplomatic demarches.

The Task Force believes that the position of Proliferation Monitoring Manager should reside in the NCPC. The Center reports to the DNI, is engaged and effective in the interagency process, and has the appropriate nonproliferation and counterproliferation coordination responsibilities. The charter for NCPC states that "it shall be the primary organization within the Intelligence Community for managing, coordinating, and integrating planning, collection, exploitation, analysis, interdiction and other activities relating to weapons of mass destruction, related delivery systems, materials and technologies, and intelligence support to United States Government efforts and policies to impede such proliferation." In pursuit of the goals of its charter, the Center has helped enable and focus military, diplomatic, and international cooperative activities and has a close relationship to those that have special authorities like CIA. NCPC draws on detailees from around the IC, the interagency, and military commands, which would facilitate the Proliferation Monitoring Manager's ability to orchestrate activities across multiple agencies, approaches, and perspectives. Elements of the Center have experience in, and a bias toward, the type of actions that this task force report advocates.

5.7. Recommendations: Improving the Tools

The "tool box" does not lack for opportunities for improvement. The Task Force recommends a number of them across the access-sense-assess-iterate spectrum.

Improving Access

The responsible agencies in the IC and DoD should rebalance existing investments and/or grow new programs in R&D to develop improved approaches for obtaining access across an array of scenarios and environments.

Expanded Sensing and Assessment

There are numerous aspects that deserve attention from each or all of the several agencies with mission responsibilities for nuclear monitoring, as follows:

- The DNI should provide direction to the leading agencies in the IC to:
  - Increase the profile, support, and integration for the OSC's Counterproliferation (CP) Program to collect and disseminate information and analysis relevant to arms control and proliferation issues;
  - Assess/adapt new and expanded collection capabilities for nuclear monitoring, especially through multi-INT integration and enhanced processing;
  - Expand the use of open source and commercial information to focus search areas and reduce demand on national collection assets so that the collection system can keep up better with the expansion of targeted areas of interest;
    - Continue/expand the augmentation of data from national intelligence collection systems with imagery and radar from commercial systems;
    - Develop and apply quality assurance methods for crowd sourcing of commercial imagery results;
  - Ensure that DoD's activities for improving global indications and warning leverage existing sources of information and capabilities, and develop the analytics to produce actionable nuclear-related threat warnings;
  - Adapt the Treaty Monitoring Manager's role to one of a Proliferation Monitoring Manager, "home based" in NCPC, with orchestration and integration responsibilities to assess both horizontal and vertical proliferation.
- DOE/NNSA, DHS/DNDO, DoD/DTRA and the IC should build upon the existing MOU to improve coordination and execution of their respective radiation detection programs to:
  - Conduct systems studies and engage operators early in development to improve transition of radiation detection advances to the field;
  - Ensure that developers and users agree in advance on system concepts, measures of success and levels of readiness for the principal technologies and operational scenarios of interest;
  - Focus new efforts on accelerating development of research with near-term payoffs, and investment in longer range technologies that can meet both technical *and* operational feasibility requirements;
  - Develop managed access to nuclear facilities and test ranges by all involved agencies at which detection technologies and operational approaches can be explored using real SNM;
  - Agree on investment strategies. There should be supporting inter-agency roadmaps to integrate efforts and focus crosstalk. Annual inter-agency reports to the NSS should be issued to track progress and enhance information flow.[36]

- DOE/NNSA, DTRA, and the IC should rebalance existing investments to grow new programs in R&D that expand activities and the supplier base to include adaptation of conventional warfighting ISR advances: e.g.,
  - Engage in planning and capabilities development, especially for data collection and fusion functions, for DoD's efforts to improve nuclear situational awareness;
  - Support transition of multi-INT fusion and exploitation tools to nuclear monitoring applications;
  - Ensure activities related to nuclear weapons and materials monitoring are guided by the "White Team" function as discussed in the previous section and Section 4;
  - Make explicit the requirements for, and improvements needed in, HUMINT, SIGINT, cyber, OSINT, etc., to support monitoring and verification.
- NSS should monitor closely and persistently the resourcing to modernize the USAEDS with the help of the VCJCS-, who should ensure its modernization is supported in AF budgets.

SECTION 6. EXPERIMENT TO ITERATE AND ADAPT: NATIONAL TESTING CAPABILITY

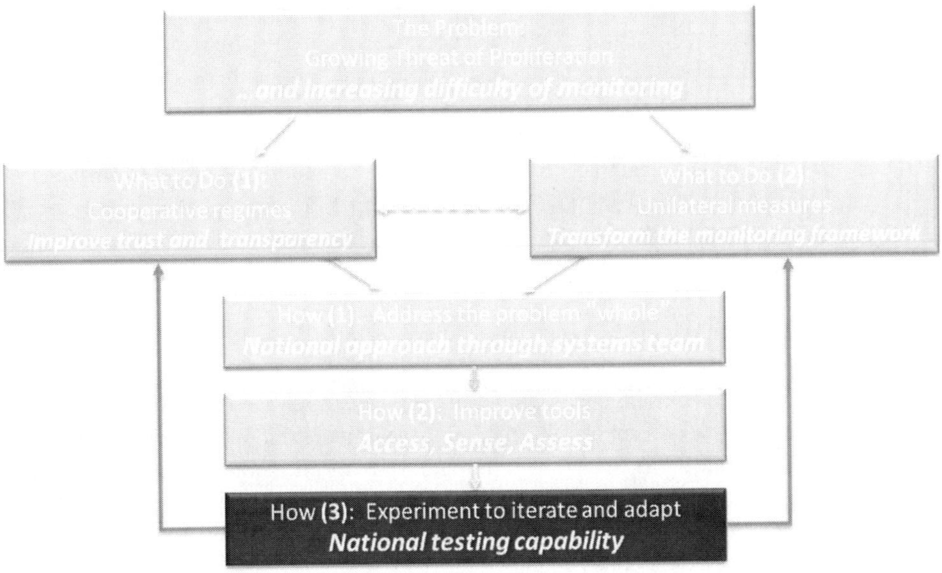

6.1. Introduction

The last of the "hows" important to comprehensive and effective proliferation monitoring is the ability to experiment with both existing and new capabilities in response to—or ideally, in anticipation of—an increasingly wider range of proliferant strategies and tactics. The Task Force addressed three inter-related topics related to experimentation: 1) a national testing capability for supporting experimentation; 2) a sample problem description of monitoring dual-capable nuclear forces (including warhead counting for arms control treaties) where the need for experimentation seems clear; and 3) use of the testing capability to develop capabilities for monitoring dual-capable systems.

6.2. National Testing Capabilities

Any argument for a national testing capability should provide answers to five important questions: (1) Why are *national* M&V testing capabilities needed? (2) What should they consist of? (3) What would be done using them? (4) Why do we think the national approach would work? (5) What should be done to get started?

6.2.1. Why Is a National Testing Capability for Monitoring and Verification Needed?

There are many inter-related reasons. The comprehensive monitoring regime the Task Force proposes is a system of systems that must work together, and is too complicated to plan/assess on paper and/or with piecemeal testing. Furthermore, experience shows that operators typically learn to use actual monitoring systems differently and often better than system design/analysis would predict. Two examples illustrate the point:

- Past experience in radiation detection operations; e.g., Operation Morning Light (the 1978 deployment to search for and recover debris from a reactor-powered Soviet RORSAT that re-entered in northern Canada); and Navy experience in detection of shipboard nuclear weapons. In both cases, operators learned to reject clutter and spurious signals by identifying patterns inherent in their own operations and/or in the larger context. For example, in Morning Light, radiation signatures from the low-grade uranium-ore deposits that are ubiquitous in Northern Canada confounded the search operations in the first days. But operators identified the source of the spurious signals, and learned that the radiation signatures from these deposits rose and declined more slowly, as the search platforms flew across the terrain, than the radiation signatures of the point-targets they were looking for, and could be sorted out on that basis;
- Recent experience in the development of new-generation tactical ISR in the Counter IED fight discussed elsewhere in this report; learning from operational experience in Iraq/Afghanistan ("on the job training"), has been coupled with iteration in experiments in various DoD testing and training facilities (e.g., the National Training Center [NTC]) to not only train deploying units but also to anticipate next steps in adversary tactics, techniques and procedures (TTPs).

The need to learn from experimentation and experience is even greater in the areas addressed by this Task Force, where an adaptive adversary and/or one with sophisticated denial and deception is likely to be involved.

The advantages of a testing capability are many and compelling. It would or could:

- Provide a focal point for planning, iterating/adapting, and operating the system of systems;
- Help integrate the technically disparate and organizationally disaggregated activities that comprise the national monitoring system;
- Better couple developers and users and provide an experimental basis for net assessment and risk management;
- Provide ground-truth performance metrics for technologies and operations, in part as a basis for future spirals;
- Stimulate sorely needed new ideas;
- Expose "subcritical" programs because planning experiments would force development of concrete requirements and CONOPs;
- Elevate attention to red-teaming of monitoring systems.

On a broader scale, it could help build international understanding and cooperation on monitoring systems. In particular, testing operations could be a vehicle for confidence building and transparency assessment and promotion, and could also contribute to dissuasion and deterrence.

6.2.2. What Would a Testing Capability Consist of?

The testing capability would have four interdependent parts: 1) the ranges and facilities (real and virtual); 2) the information/data management systems (data flows from and to sensors, information flows within the analysis/fusion systems, and command/control data/information flows); 3) a standing "White Team"; and 4) a standing Red Teaming activity.

The ranges and facilities would consist of a distributed, netted set of:

- Operating real-world facilities (e.g. PANTEX, a power reactor, port of New York, SFO, the URENCO enrichment plant);
- Real test facilities (e.g. NTC, Nevada National Security Site [NNSS]) with space for large-scale operations and/or use of radioactive materials;
- Simulated facilities/operations.

This set of facilities and networks would eventually cover the full range of nuclear threat and treaty-related capabilities and activities, including: labs, production, industrial base, distributed weapons and materials storage, deployment, distributed field operations. The set would also be broad enough to bring into play "patterns of life" in peace, crisis, and perhaps war. The facilities should include remote areas, areas like "the South Bronx", air, and at-sea operations. There should be the capability to use real SNM. Some facilities and ranges might be overseas, including facilities that would be the actual subject of monitoring in future treaties.

Information and data flows are the second crucial part of the testing capability because they comprise the essence of monitoring. The monitoring system itself is about information; monitoring the data/information flows in treaty-partner and/or adversary systems and operations is a crucial function of the monitoring system. The testing capability must thus include both blue and (hypothesized actual or simulated) red information/data management systems and networks. Such information/data systems—real or simulated—can be very expensive, and their development and iteration very time-consuming. Fortunately, rapid prototyping, testing capabilities, and iteration for such data management systems have been developed for other applications. It appears possible to readily adapt such testing for the information management aspects of the nuclear monitoring testing capability. Moreover, the testing capability systems can evolve to become the actual operational systems.

The third integral part of the testing capability would be a standing analytical "White Team" that would use the scenario-generator and scenario-to-asset planning approach (see Section 4) to guide exercises/experiments, assess implications of testing and training results, and identify needed changes/improvements for the next cycle. It would provide a simulation capability and translate test results to deployed monitoring operations. The information management aspects of the White Team activities would eventually become part of the real world ops center for the comprehensive nuclear monitoring system of systems.

The fourth essential element of the testing capability would be a standing Red Teaming activity. Red teaming would include active R&D on means by which adversaries could (or do) evade threat-assessment monitoring and by which treaty partners might evade treaties and treaty-monitoring, including by deception and denial. A key step would be to rebuild a national nuclear threat assessment capability, principally but not exclusively with the National Labs, that anticipates "nth group" weapon designs, development, and production approaches (as discussed in Section 3). The services and certain combatant commands (COCOMs) could explore ways in which nth-countries might deploy, operate, and employ nuclear and dual-capable weapon systems and forces. The White Team would referee red-teaming, for example by specifying the levels of technology, or degree of knowledge of U.S. systems, that might be attributed to various actual or simulated adversaries.

6.2.3. What Would Be Done Using the Testing Capability?

The testing capability would eventually cover the full range of experiment, test, demonstrate, exercise, and train, through multiple cycles of learn-iterate-adapt. The testing capability could be used to explore very specific detailed topics, like improved CONOPS for a particular type of advanced sensor, and very general questions, like the interplay between transparency and stability. Each experiment would use a few or many of the types of facilities and capabilities that are part of the testing capability. The exercises, tests, etc., would explore the full scope of current and possible future monitoring applications: nonproliferation, counter-proliferation, countering nuclear terrorism, IAEA monitoring applications, Comprehensive Nuclear-Test-Ban Treaty (CTBT) monitoring, cooperative threat reduction, routine and challenge on-site inspections, negotiated over-flights (Open Skies done right – see Section 2), and confidence- building measures.

The exercises and experiments would be sponsored by all relevant USG agencies, occasionally singly but more often jointly. In certain important cases there would be international participation, partly for confidence-building purposes or ideally, as an integral part of the phased strategy for cooperative regimes discussed in Section 2. There would be red- and blue- teaming, in multiple forms.

6.2.4. Why Does the Task Force Think the Testing Capability Would Work?

Almost all successful developments of major systems of systems have used testing capability development and iteration. DoD in particular has a long history of using testing and experimentation, augmented by simulation, to develop successful systems. The near real time experiment-iterate-adapt cycle for countering IEDs simultaneously developed technical capabilities (e.g., ODIN & MAD DAWG) with CONOPs. Earlier, during the Cold War, there were key developments in which test-beds used in conjunction with operational experience led to important improvements in theater weapons, forces, and their CONOPS. These included:

- "Shockwave" in the North Atlantic Treaty Organization (NATO) Central Region during the 1970s and 1980s (see the next section of this Section for a more complete description);
- SORAK (N.E. Asian theater)[37];
- Pershing IA and ground launched cruise missile (GLCM) endurance experimentation during the 1980s[38];
- Horizontal and vertical dispersion of dual-capable aircraft (DCA) to improve theater nuclear force (TNF)-DCA force endurance.

Major force component assessments and requirements have also been developed using test- bed approaches, including the Navy's submarine survivability program and the Air-Land battle command robustness.

The nonproliferation and arms control community is not a stranger to such approaches, although systems integration and experimentation are not as common. For example, Sandia's Technical On-Site Inspection (TOSI) site was used effectively to develop on-site inspection approaches.

6.2.5. What Should Be Done to Get Started?

The first application of the testing capability should be one for which progress would address a major shortfall in current capabilities and possibly open the door for new arms control agreements. Monitoring dual-capable/TNF weapon systems was selected as a leading example for these reasons. The challenges for monitoring TNF systems are discussed in the next section, followed by ideas for getting started on learning how to monitor them using the testing capability in Section 6.4.

6.3. Monitoring Dual-Capable Nuclear Forces, Including Warhead Counting for Arms Control Treaties

Monitoring dual-capable (nuclear and conventional) systems is of interest in its own right, but it also epitomizes other monitoring problems. It illustrates the synergies among the four elements of the Task Force's basic 2x2 matrix of Figure 1-2, which encompasses the full scope of the monitoring regime; i.e., monitoring for the purpose of treaty-compliance and monitoring for threat assessment, within the context of both negotiated and/or cooperative monitoring agreements and non-negotiated/unilateral monitoring.

All nations that have nuclear weapons, except for the UK, have dual-capable (DC) weapons, and many of these nations, including recent proliferators, have *only* dual-capable nuclear systems. It is likely that most potential future proliferators (Iran, etc.) will use dual-capable platforms for their initial, if not longer term, nuclear capability.

There is a wide range of dual-capable weapon -system types:

- Air forces: gravity bombs, air-launched cruise missiles, air-to-air missiles, defense suppression missiles.
- Ground forces: anti-armor infantry weapons, tube artillery, surface-to-surface missiles, air defense missiles, demolition munitions; some are dual-role (e.g., surface-to-air and surface-to-surface).
- Naval forces: surface-to-surface missiles, air defense missiles, ASW weapons, torpedoes
- Special forces weapons

We have also seen that some nuclear warheads may be dual-use (i.e., strategic and TNF), adaptable both ways.

It is important to take into account that most proliferators will rely on "hiddenness" and uncertainty in general, for their nuclear systems survivability. The survivability provided by hiddenness and uncertainty tends to enhance crisis stability but exacerbates the difficulty of monitoring and verification for both arms control and threat assessment. The net effect may be that the risk of miscalculation by current or potential future adversaries increases, and overall strategic stability over the long term decreases. How transparency, certainty and uncertainty, and stability interact, for good or ill, is an important issue that deserves more systematic thought than it has received to date. Dual capability is an integral part of this question, because of the uncertainty about whether any given delivery platform may be armed with, or destined to be armed with, a nuclear weapon.

To be effective, monitoring of dual-capable forces must cover the full spectrum of activities: design, production, storage, and deployment; during both peacetime and crisis/war operations; for both nuclear warheads and DC systems. Understanding this complete set of activities—i.e., the "pattern of life" for dual capable systems—illuminates each individual part, including parts that may be relevant to treaties, but also paints a much broader picture of nuclear capability. An increasingly serious problem is that U.S. de-emphasis of dual-capable systems narrows our frame of reference for monitoring and verification of others. The result is a limited range of opportunities to test monitoring systems against our own dual-capable systems, and the roster of people who understand the possible patterns of life of DC systems, based on their own experience with them, is shrinking rapidly. It is a central theme of this report that there are synergies that should be amplified and exploited between monitoring for arms control purposes and for threat assessment, and between negotiated/cooperative monitoring and monitoring means that are not negotiated and not explicitly cooperative. These synergies are illustrated, in more specific terms, in the case of monitoring for dual-capable systems, in Figure 6-1.

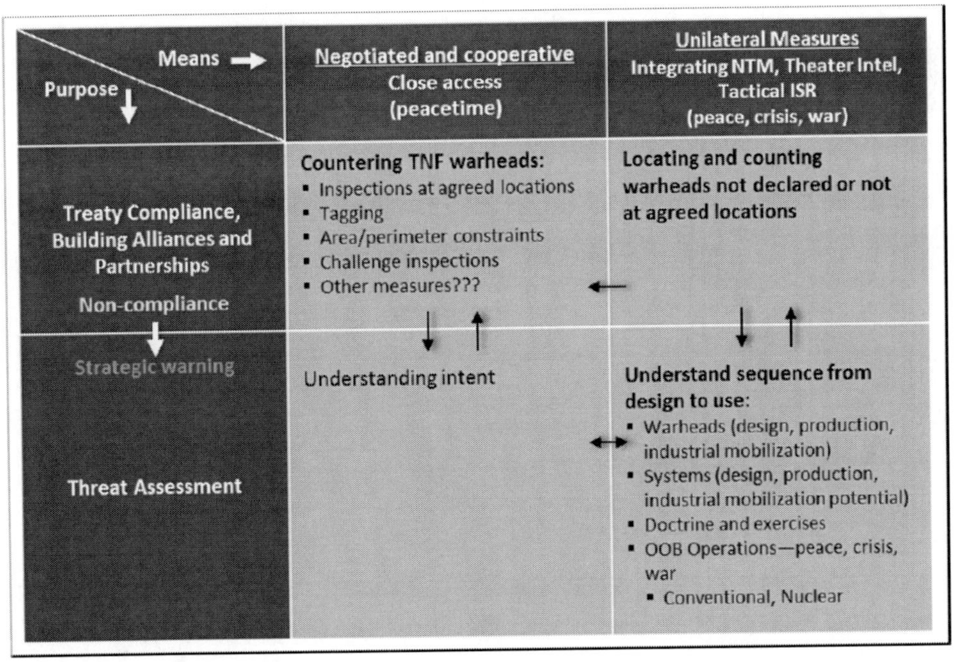

Figure 6-1. Dual Capable (DC) Theater Nuclear Forces (TNF) - An Example to Illustrate Synergies in our Approach.

One can observe that:

- Knowledge of warhead inventories gained from treaty monitoring contributes to threat assessment. Verified or even suspected non-compliance with treaty limits would represent a form of strategic warning in the threat assessment context.
- The objective of threat assessment is to know everything relevant about the force structure, order of battle, development life-cycle and operational patterns of life in peace, crisis, and war, for both warheads and systems. Much of this knowledge would also be needed in order to assess whether warheads were being withheld from the treaty-prescribed counting-process.
- Negotiated limits on numbers of warheads narrow the range of uncertainties in understanding of the overall TNF/DC posture for threat assessment purposes.

Some of the monitoring means that might be involved in the space of Figure 6- are shown in Figure 6-, a version of the same basic framework.

In the upper left corner of this graphic, the Task Force posited five possible elements of a treaty involving counting TNF warheads. On the right side, a few candidate monitoring means for both complementing the treaty-allowed monitoring means and for monitoring for threat assessment are listed. While the Figure suggests a clear distinction between methods applicable to Treaty Compliance (top row) and Threat Assessment (bottom row), in reality, the methods are applicable across both.

Means → / Purpose ↓	Negotiated and cooperative Close access (peacetime)	Unilateral Measures Integrating NTM, Theater Intel, Tactical ISR (peace, crisis, war)
Treaty Compliance, Building Alliances and Partnerships Non-compliance	▪ Count declared w/hs at declared sites ▪ Verify absence at non-allowed sites ▪ Monitor agreed boundaries ▪ Challenge inspections ▪ Specified overflights	▪ NTM/NMM ▪ Shockwave-like operations ▪ ODIN-like systems ▪ Covert UGS ▪ Covert tagging ▪ Crowd-sourcing ▪ Attaches ▪ ...
↓ Strategic warning Threat Assessment		▪ Haystack-like analysis ▪ "Patterns of life" ▪ Nth Country's nuclear designs

Figure 6-2. Dual Capable (DC) Theater Nuclear Forces (TNF) -Some Specific M&V Approaches for Monitoring.

"Shockwave" (mentioned in the previous section) was a comprehensive, successful, TNF-related threat-monitoring/understanding effort that ran from about 1978 to about 1985. It is discussed here as an effective antecedent of future, comprehensive TNF threat-monitoring systems which would benefit from improved technologies and possibly from negotiated limits and monitoring. The purpose of Shockwave was to understand improvements in the ability of the Soviet/Warsaw Pact (WP) to execute a swift, successful campaign inside U.S. and NATO timelines for nuclear employment, and thus to aid selection among U.S. and NATO options to counter those improvements. It used a wide range of U.S. national, theater, and tactical monitoring means, from both the Army and Air Force, as well as similar Allied means where appropriate. The Defense Advanced Research Projects Agency (DARPA) and the Defense Nuclear Agency (DNA) were major players in developing new monitoring capabilities. Exercises using NATO forces were used to elicit responses in later WP exercises that could be observed by Shockwave assets. The effort was led by successive SACEURs, with integration at both Supreme Headquarters Allied Powers Europe (SHAPE) and United States European Command (EUCOM) headquarters. Over the 7-8 years it was run, Shockwave was highly successful. In fact, it serves as a premier example of how persistent, comprehensive monitoring for threat-assessment purposes can pay off.

Shockwave involved development and use of persistent (for its time), relatively wide-area (for its time), aircraft-based, multi-sensor, monitoring systems, combined with many other approaches to gaining understanding. There is a direct evolutionary path from those systems to the kinds of systems used, recently and currently, for more-persistent, wider-area monitoring to suppress the IED threat in Iraq and Afghanistan. One such recent system was ODIN. ODIN- like systems, in turn, could be adapted and extended for a future, Shockwave-like effort for monitoring dual-capable systems (as well as for other purposes, including wide-area search for "loose nukes"). Integrated with such systems could be a wide range of other things including covert unattended ground sensors, covert tagging, and crowd-sourcing for gaining patterns-of-life information. The ODIN concept has been conceptually adapted for nuclear monitoring; the operational concept demands close access for detection consistent with the discussion on radiation detection in Section 5.

We have focused here on monitoring of dual-capable systems both because they are relevant in their own right, for future treaties and for threat assessment, and because other M&V problems share aspects of the DC/TNF M&V problem. For example:

- Monitoring the IAEA Additional Protocol, which would allow access to undeclared nuclear facilities, is similar to the problem of finding undeclared warheads;
- Awareness of the early stages of proliferation and monitoring weapon production are similar to understanding the dual-capable weapon system development and deployment life cycle.

6.4. Use of the Testing Capability for the TNF-DC Problem

The ODIN system of systems became effective for IED suppression because its users were forced to learn from experience in the real world, including from many initial failures where people died. Nuclear monitoring systems for both treaty-monitoring and threat-assessment will not have the same plethora of daily, high-stakes, real-world events from which to learn. An essential part of developing and iterating these systems toward success must be providing frequent learning experiences using the test-bed approach we discussed above. We close by returning to that subject.

As introduced earlier in this Section, the first use of the national M&V test-bed should be for both arms-control and threat-assessment monitoring of dual-capable systems. The key is to start small in both the scope of the facilities/ranges involved and the complexity of the scenarios, and then grow in both dimensions as national monitoring capabilities grow and need to be integrated. The goal should also be to start soon—the problem space is too complex to be designed perfectly a priori.

As a first step, DTRA should analyze use of wide-area, persistent ISR/reconnaissance, surveillance, and target acquisition (RSTA) for monitoring of dual-capable systems to support both treaty-monitoring and threat-assessment. Pros and cons should be developed comparing various applications of this first step, including for example, future warhead-counting treaties, assessment of Russian dual-capable systems in general, and assessment of China's DC systems along the Taiwan Straits.

Based on that analysis, the next step should be to plan, jointly among DoD, DOE, and the IC, a first, relatively simple experiment in detail; for example, using an ODIN-like system to find a single dual-capable platform operating in "move-and-hide" mode, at NTC and/or NNSS. (The latter would allow experimentation with real SNM signatures.) This would be a relatively low cost initial effort, but it would get started on an important problem. Even this first small step should involve red-teaming, and should be planned by the White-Team using the scenario- generator described in Section 4. Various sensor mixes and adaptive sensor/platform CONOPs, posited by a "solutions team" also described in Section 4, should be explored. The product should be a well-understood design for one component of the future nuclear monitoring system of systems, along with options for operating that component.

Following success with this single-focused monitoring application, the experimentation campaign should add two, three, and more elements of the "pattern of life" for the red nuclear enterprise. The results should not only inform CONOPs development, but also identify intelligence gaps where collection assets might be focused.

For all of the above, it will be important to partner with at least one geographical COCOM, and to be guided by a steering group of military and technical experts who have made such undertakings a success in the past.

6.5. Recommendations: National Monitoring Testing Capability

The Task Force believes that even with substantial investment in all other elements of its proposed "whats" and "hows", the U.S. will likely see only marginal improvement in monitoring capabilities without the national testing capability. While the idea of "testing capability" could suggest an expensive undertaking, the Task Force has been careful, as it has in other parts of this report, to make recommendations that build upon existing capabilities or shift investment priorities within existing portfolios. In the case of the testing capability, the communities should make maximum use of existing capabilities in ranges, equipment, and development programs (both hardware and software) that already exist. It is the integration and application of such capabilities to the "paradigm shift" in monitoring that is important for learning and making progress in addressing the proliferation problem. In fact, the Task Force believes that its "get started" plan can be undertaken without substantial new investment outside of intellectual capital. To that end, the Task Force recommends that:

- DoD/DTRA, in partnership with DOE/NNSA and the IC, develop a comprehensive "iterate and adapt" national testing capability which:
  - Provides a focal point for planning, iterating/adapting, and operating the system of systems;
  - Helps integrate technically disparate and organizationally disaggregated activities.
- The testing capability should include the following elements:
  - Red and blue teaming, in multiple forms, to include engagement of allies in various red and blue roles;
  - Realistic physical and virtual environments;
  - White Team (as recommended in Section 4); i.e., a strong analytical component to assess the implications of test results and feedback into the cycle needed changes/improvements.
  - Support from DOE, working with NCPC, to rebuild a national nuclear threat technical assessment capability, principally but not exclusively with the National Labs, that anticipates "nth group" designs, effects, employment and deployment approaches and observables.

SECTION 7. SUMMARY AND CONCLUSION

The Task Force's mandate was broad, as specified in the Terms of Reference. As such, many topics did not receive the in-depth look that would help executing organizations with more explicit recommendations on priorities and get started actions. For example, the Task Force fell short in providing explicit recommendations for:

- Defining specific steps for the long term cooperative engagement plan;
- Assessing more thoroughly the opportunities and "dry holes" in applying conventional warfighting ISR capabilities;
- Assessing the current and desired information flows and integration for early detection of proliferation;
- Recommending in greater detail improvements in collection and analysis carried out by the IC as part of the overall monitoring plan;
- Advancing materials monitoring and safeguarding technologies;
- Creating and sustaining the "White Team";
- Developing a national RDT&E program for improving and fielding radiation detection systems;
- Defining the specific requirements and implementation path beyond the initial step recommended in the report for a national testing capability.

That said, the Task Force believes the comprehensive look it provided is both needed and timely. The topic has not been previously addressed in anticipation of the nuclear monitoring and verification demands that the nation is likely to face in the future. While difficult to predict, that future seems poised to present challenges unlike any faced with the experience base derived from historic nonproliferation and arms control treaty regimes. This is due to a number of reasons highlighted throughout the report, but what to do about it forced the Task Force to ask whether simply doing better at what we already know how to do would be sufficient. The answer the Task Force came to was "no". Instead a combination of evolution and innovation is called for.

In that context, the Task Force admonishes leadership not to repeat shortcomings of the past where monitoring technology and verification efforts were tied to the treaty or the proliferation problem immediately at hand, and then declined when that problem had been addressed or had diminished in perceived importance. The problem of managing the global nuclear environment for stability will be with us for a very long time. The best bet for making progress is a sustained effort in which experienced and competent professionals can devote their careers to the quest and pass on their wisdom to successive generations.

Success in this long and important effort is not guaranteed, considering the difficulty of the future monitoring challenges. But progress can be made, building on successes of the past against equally daunting problems, and on experience with prudent hedging strategies and programs.

Appendix A. Unabridged Description of a Proposed M&V Analytical Methodology

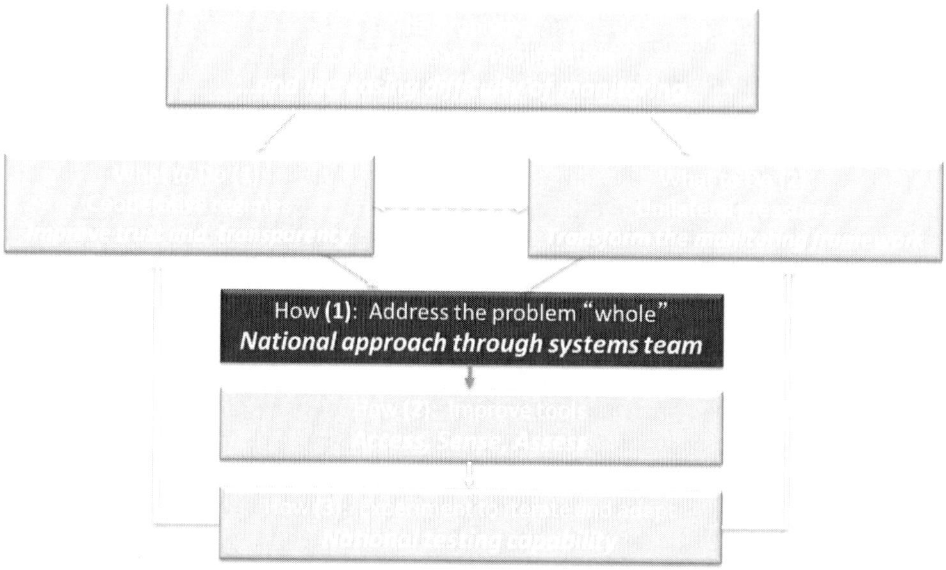

Section 4 presented a candidate analytical methodology for prioritizing among the many operational and acquisition options to advance the nation's monitoring and verification capabilities. A more complete description is provided in this appendix.

A.1. Elements of an M&V Analytical Methodology

There are several key elements proposed as a necessary part of an enduring M&V analytical methodology, as laid out in Figure A-1.

- First, there is the "problem space", which consists of a set of frameworks that describe the M&V challenge independent of any proposed solutions. This is a critical component in the proposed approach, as it provides the basis that will serve as a common foundational understanding for solutions to be built upon and assessed for fitness.
- Second, there exists a "solution space", in which proposed capability architectures are crafted. These architectures are collections of technologies, operations, and capabilities that work together to accomplish the goals and objectives identified through examination of the problem space.
- Third, there must be a Bridging Methodology which allows for the "back and forth" between requirements and objectives that any solution must meet, as dictated by the frameworks that describe the problem, and architecture features and assessments of the solution space that are mapped into the analyst's understanding of the problem itself.

- Finally, there must be a rigorous and repeatable portfolio decision methodology that allows for the values of the decision maker and his/her organization to be incorporated, and for defensible, tractable decisions to be made.

A common issue in the development of analytical frameworks has been the tendency to frame the problem and solution together. The conflation of "problem space" and "solution space" brings with it several consequences. First, it tends to promote metrics and assessments that maximize elemental solution parameters such as detector performance, rather than system performance against the actual threat space, by assuming that they are proximate. Second, it tends to lead analysts to fixate on a narrower problem (such as detector performance) rather than assessing those issues in the context of the larger whole. Third, it does not enable a common understanding of the problem space itself, allowing solutions to be framed in whatever manner the solution proponent deems suitable. The entire set of symptoms run counter to enabling the decision maker to achieve his/her ultimate goal: rendering and defending a provably effective investment strategy for M&V capability development. Separating the "problem space" from the "solution space" and analyzing them independently can therefore yield benefits by eliminating or mitigating these issues.

To illustrate the point, consider that detection systems have often been placed at the forefront of the national strategy for detecting the illicit movement of nuclear threats in proliferation regimes. While undoubtedly detectors play a role in the solution, the bulk of current analytical activities are focused on detector system performance. As a result, the M&V problem of detecting illicit movements is often miscast implicitly as a detector problem. This can lead to the line of thinking that more detectors with better detector performance parameters must logically provide reduced risk – a statement that may or may not be true. More importantly, other options that do not hinge on detector deployments may in fact provide more cost effective mechanisms for risk reductions.

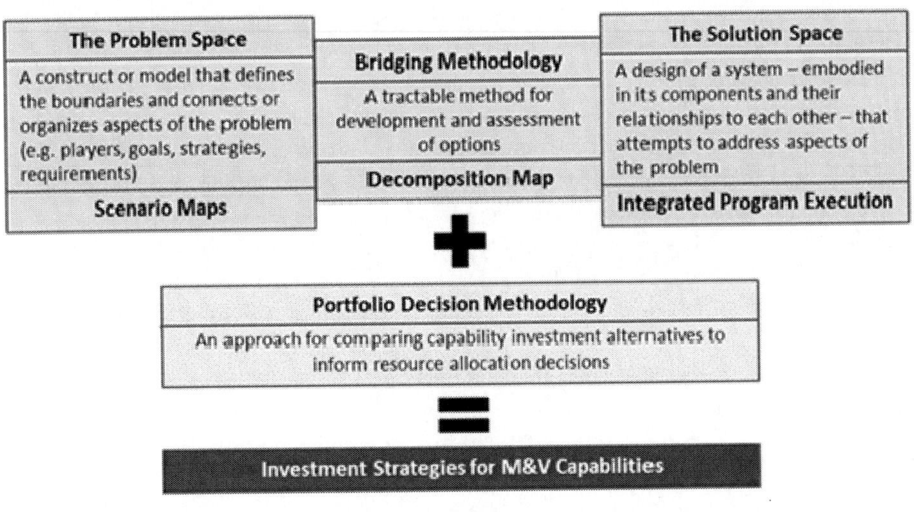

Figure A-1. Elements of an M&V Analytical Methodology.

A.2. Proposed Problem Space Description

While there are several possible frameworks for describing the M&V problem space, the Task Force chose to use one based on scenarios. This section describes the scenario framework and its application for describing the M&V problem space. For the purposes of this effort, the following definition of scenario is used:

> A scenario is an evolution of the world through a series of incremental events from its current status towards an outcome of interest that is specified by the analyst[39].

Utilizing scenarios for the evaluation of potential solution sets to complex problem spaces is by no means a new idea. Neither is using scenario sets as an organizing framework for defensive architecture development across a disparate set of stakeholders. In a report that attempted to pave the way for improving U.S. capabilities to combat biological terrorism, Danzig suggested that a common set of planning scenarios be adopted by the community concerned with bio-terrorism, along with a set of metrics, and proposed four scenarios for consideration.[40] This approach had several attractive potential benefits. First, it laid out scenarios that were wholly distinct from any proposed solutions, ownership, or mission space, but instead focused on a concise model of the problem of bioterrorism itself. Second, it began to establish a consistent framework for the development and use of metrics and measures of performance. Finally, it highlighted the stewardship of problem definition as a necessary and important element in the set of capabilities to combat the problem itself.

Scenario based planning approaches can also have shortcomings if not appropriately implemented. For example, while useful for highlighting specific trade-offs, a limited set of scenarios can generally only cover a small subset of the variables that an analyst might want to explore, thereby artificially narrowing the problem space. Utilizing a wide set of end-to-end scenarios can become cumbersome and difficult to assess systematically. Additionally, conclusions drawn from a narrowly defined set of specific scenarios can be fragile to "what-if" challenges, where the scenario details beyond those explored in the analysis are modulated, casting uncertainty on the results and raising the risk of invalidation. Scenario authors also run the risk of fixating on "favorite" scenarios, or those that mirror their own preferences and biases, which can prevent systematic and objective thinking. Finally, a non-systematic scenario approach can exacerbate the problem of mapping scenario details and analysis to a structured decision methodology.

Rather than utilizing a small set of well-defined scenarios with the risks and shortcomings noted above, the Task Force developed a scenario framework that attempts to encompass a large portion of the potential scenario space in order to address the totality of M&V problem complexity. A scenario framework for scenario generation and analysis is desirable primarily because it enables the examination of a family of scenarios, rather than a small set of independent and specific scenarios. It also provides a systematic method for decomposing scenarios into discrete nodes and linkages, and capturing the interdependencies between individual scenarios. Finally, it lays the foundation for a bridging methodology, or systematic mapping between the problem space and solutions space, that enables increased traceability between planner objectives and solution performance.

In constructing an example scenario framework, the study team began with a simple premise: a scenario is a linear series of events can be broken down into discrete nodes and associated linkages, and systematically analyzed, node by node. **Error! Reference source not found.** shows a simple sequence of nodes within the framework to illustrate these concepts. In this example, the failure of a nation's NW security structure (e.g. compromise of storage site security, material stolen by an insider) begets an evolution of subsequent scenario actions. Such a failure could lead to the theft or loss of a weapon from the custody of a nuclear weapon state. This theft could lead to the transition of a weapon to a non-nuclear weapon state, thereby proliferating not only weapons possession, but also technology and knowledge.

Each scenario node could play an elemental role in many other scenarios – e.g., the theft of a weapon could be a common step in several distinct scenarios. The nodes are therefore linked in a network of connections that becomes a framework for systematically developing and analyzing an entire family of scenarios in the M&V problem space. Figure A-2 illustrates the complete scenario framework, rotated to fit the page, while Appendix B provides a larger view of the framework.

As shown in Figure A-3, scenarios generally evolve as one traces from top to bottom, following the directed graph towards consequences. The left side demonstrates scenarios that can lead to movement of weapons themselves from NWS to other actors. The center details scenarios that can lead to the proliferation of NW technology and knowledge, and the right hand side covers compromise of weapons usable nuclear material (WUNM) and associated technology. The long vertical chain in the middle-right illustrates the evolution of non-state actors towards unconventional attack with nuclear weapons. Consequences in the scenario framework include blackmail, attack, or simply the emergence of a new nuclear-armed state.

Within the scenario framework, blue-colored nodes indicate "macro" nodes, or nodes that are easily decomposed into a common set of sub-nodes. For the purposes of this work, the only set of sub-nodes explored is the "movement sub-space." This sub-space corresponds to movement through the origin region, across international boundaries, through global commons, and into the target region. The movement sub-space is laid out in the blue box on the right hand side of the framework diagram.

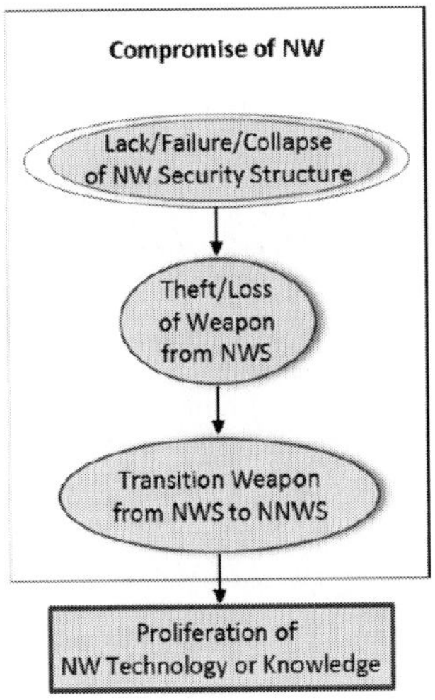

Figure A-2. Example Node Sequence.

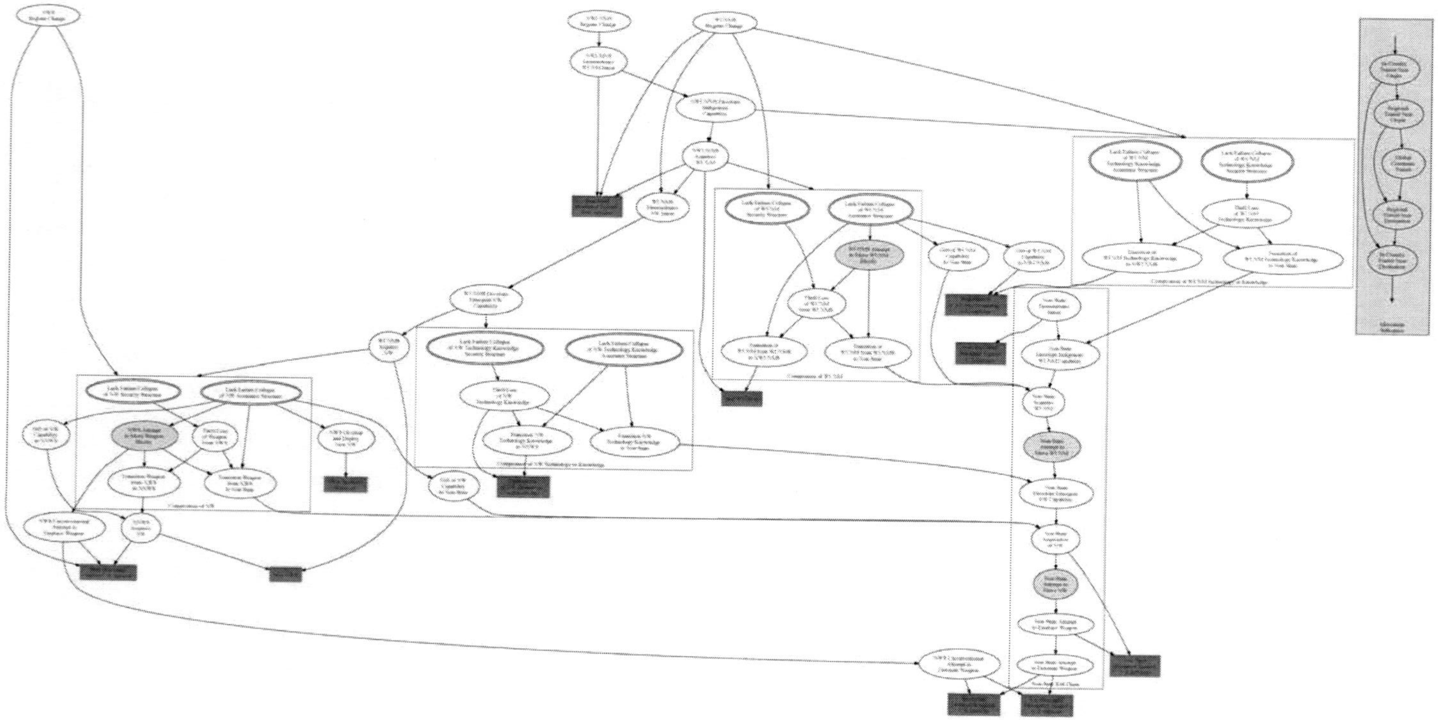

Figure A-2. Proposed Scenario Framework.

Real-world events could easily be interpreted as a signal that a scenario in the framework is playing out, although there is actually no nefarious activity. It is very important to consider these degenerate scenarios, in addition to actual threat scenarios, when analyzing potential solutions as it is likely the bulk of actual signals and information will be from situations that are not actually leading to nuclear attacks. For example, the nuisance alarm problem in radiation detectors that are monitoring commerce streams plays a large role in the overall cost, effectiveness, and operation of those systems. Operational and analytical understanding of such false positives leads to an understanding of the full suite of information and analysis required for high confidence detection.

A.3. Analysis within the Scenario Framework

Analytically, the scenario framework serves several roles. Most importantly, it exists as a common frame of reference for describing the M&V problem space. Narrower problem definitions, metrics, and objectives can be derived where appropriate through decomposition (discussed more in Section A.5). Scenarios for analysis can be generated by stringing sequences of nodes together. Any starting point and ending point can be selected, and a path through the network selected to connect them. From that string of nodes, a more complete narrative can be constructed.

The scenario framework can also allow for greater and more complete coverage in the design and analysis of solution architectures. A large family of scenarios can be analyzed by examining all nodes systematically node-by-node independent of end-to-end scenarios. An analyst can consider solution architectures that combat adversary success within a single node, and consider the collective impact it has on the complete scenario space by examining both upstream and downstream nodes. Additionally, tradeoffs between architectures designed for different nodes can be compared in an end-to-end system performance sense. This will aid the assessment of the complete set of architecture components within a portfolio of defensive measures and allow for complex trades to be made. While the scenario framework provides the structure for this kind of analysis, there is still the challenge of developing end-to-end metrics that are solution independent and common among all nodes. Further consideration of this issue is given in Section A.8.

Finally, it should be noted that the scenario framework is not proposed to be the sole description of the problem space for the M&V mission. Rather, it is one useful construct for describing the problem space that lends itself to the analysis that is discussed in this Section. As is often the case with large complex problems that have a myriad of stakeholders who might have competing objectives, there are likely many representations of the problem space that are all germane to evaluating potential solutions. While it is recommended that the scenario framework be considered as a unifying representation of the problem space for analysts and decision makers, it is also recognized that other representations exist and are useful. Such work should continue, but the need for integrating them into a common set of problem space models remains. Likewise it is recommended that the scenario framework be periodically reviewed and updated as appropriate based on real world experience and additional analytical studies.

A.4. Bridging Methodologies

While the scenario framework proposed in Section A.2 can provide a common frame for understanding the M&V problem space, any problem framework must be connected to potential solution options in order to be ultimately useful to capability investment decision makers, or policy makers and treaty negotiators. The method used to connect the problem and solution spaces is called a bridging methodology for the purposes of this report. A bridging methodology approach allows for a breakdown and prioritization of goals and objectives in the problem space into requirements and metrics for potential solution architectures. It will also allow for the systematic aggregation of performance assessments and analyses into an overall picture of monitoring and verification architecture performance.

A.5. Proposed Decomposition Map Approach

The bridging method proposed by the Task Force is a decomposition approach that maps between problem space descriptions and prospective solution architect elements. This subsection describes the decomposition approach in general; the next one provides an example to illustrate its application.

As envisioned, the proposed approach begins with the selection of any node in the scenario framework discussed in Section A.2. The selected node is decomposed into any sub-nodes required to add appropriate fidelity or resolution to the analysis. A system of decomposition layers is then constructed beneath the scenario nodes. Those four layers are:

1. **Strategic Capability Areas** – This layer, while arbitrary, provides a convenient organizational structure when considering the universe of potential capability investments. As defined in this report, the strategic capability areas center on core elements of the mission space associated with reducing risk.
2. **Functional Objectives** – Within each Strategic Capability Area, several functional objectives can be articulated. These objectives are intended to capture high-level operational objectives that must be achieved. The articulation of these objectives must be performed by the decision maker, as there is no universal set. The metrics used to assess performance against those objectives must be derived by the analysts from the articulation of risk in the problem space.
3. **Tasks** – Each functional objective can be further decomposed into a set of tasks. The tasks themselves are part of prospective solution architecture – i.e., tasks, just like objectives, are not universally defined, but proposed as part of a solution option. Each task defines a specific component of a functional objective, to be accomplished through the application of assets. The measures of performance used to assess the performance of a task are formulated derivatively from the metrics used to assess performance against the functional objectives.
4. **Assets** – Each task is accomplished through the employment of assets. Assets can include hardware, platforms, people, training, concepts of operations, and programs – essentially any capability that can be specifically invested in. Requirements and metrics for assets are established derivatively from the tasks. The mix of appropriate

assets for consideration is dependent on the tasks proposed as part of prospective solution architectures.

As assessments of performance of existing or proposed solution architectures and components are completed through analytical work, the results must be first cast in the framework of the decomposition map, and then captured as part of the overall solution architecture assessment. Areas of disagreement in assessments, or those areas that are not well covered (or over-covered) by current analytical results can be readily identified. In this way, a decomposition map can also serve to guide the overall effort of an M&V analytical capability, such as that described in Section A.6.

Each node in the scenario framework proposed in Section A.2 will have at least one unique decomposition map associated with it in a fully formed analytical effort. There are only a limited number of unique investible assets that may be incorporated in prospective solution architectures, and many assets are likely applicable to several different functional objectives and tasks. These observations imply that most assets will aggregate requirements from multiple scenario nodes, strategic capability areas, functional objectives, and tasks. Assets must be assessed against each set, and synergies may be identified and leveraged when designing solution architectures.

Optimistically, the same asset may have sufficient performance and applicability across multiple tasks, functional objectives, strategic capability areas, and scenario nodes. This implies that if a consistent accounting of performance and benefits is made, through a systematic methodology as proposed here, then benefits across the problem space can be better understood, and investments in one area may have the effect of reducing risk in areas outside the scenario node currently being investigated. In this way, one might imagine a fully executed analytical model of the M&V mission areas as a large multi-layered network – on one face exists the problem description (i.e. the scenario framework), and on the opposite face proposed solution architectures in the form of a set of investible assets. The layers in between are highly connected and consist of the layers articulated in the description of the bridging methodology described above.

A.6. Decomposition Map Example

The overall decomposition approach described in Section A.5 may be best described through the example illustrated in Figure A-3. The example decomposition map begins with a focus on the scenario node “Non-State Attempt to Move Weapon”. The scenario node is decomposed into more specific scenario sub-nodes; in this case the analyst is concerned with “In Country Movement” of a nuclear weapon or asset.

The next step of decomposition moves from the problem description (the red shaded area in the decomposition map) towards the articulation of a solution architecture (the blue shaded area in the decomposition map). The scenario sub-nodes are decomposed into strategic capability areas dividing the problem space into discrete mission areas – in this case border monitoring is highlighted. In addition to other functional objectives, it may be desirable as part of border monitoring to control a perimeter around an area of interest, preventing nuclear materials or devices of interest from moving across that perimeter (it may also be desirable to allow innocent traffic to cross the perimeter, as in border control scenarios). Additionally, the capability to search small areas thoroughly, and search large areas efficiently may be important functional objectives.

The functional objective, "Perimeter Control" can be further decomposed into a set of tasks that must be accomplished in order fulfill that objective. These tasks and the assets associated with them represent a proposed solution architecture for accomplishing the objectives specified. For example, the perimeter must be established and maintained. Additionally, items that pass through that perimeter must be inspected to ensure that no nuclear contraband is passing through the perimeter. A secure zone between inspection points must be maintained to ensure that no nuclear contraband is passing around those inspection points. (Additional tasks may need to be performed, but are not included here.)

In this example, it is proposed that radiation detectors, in addition to persistent surveillance assets might be used to help establish a perimeter through monitoring and surveillance of perimeter zones. Boots on the ground and vehicles are proposed for enforcing perimeter incursions presenting a show of force along the perimeter itself. Similarly, additional assets can be laid out and their relationships to mission objectives captured. The requirements and metrics by which proposed assets are to be assessed have been derivatively defined from the previous layer. As assessments are made and analytical results obtained, those are aggregated into performance assessments of higher and higher layers.

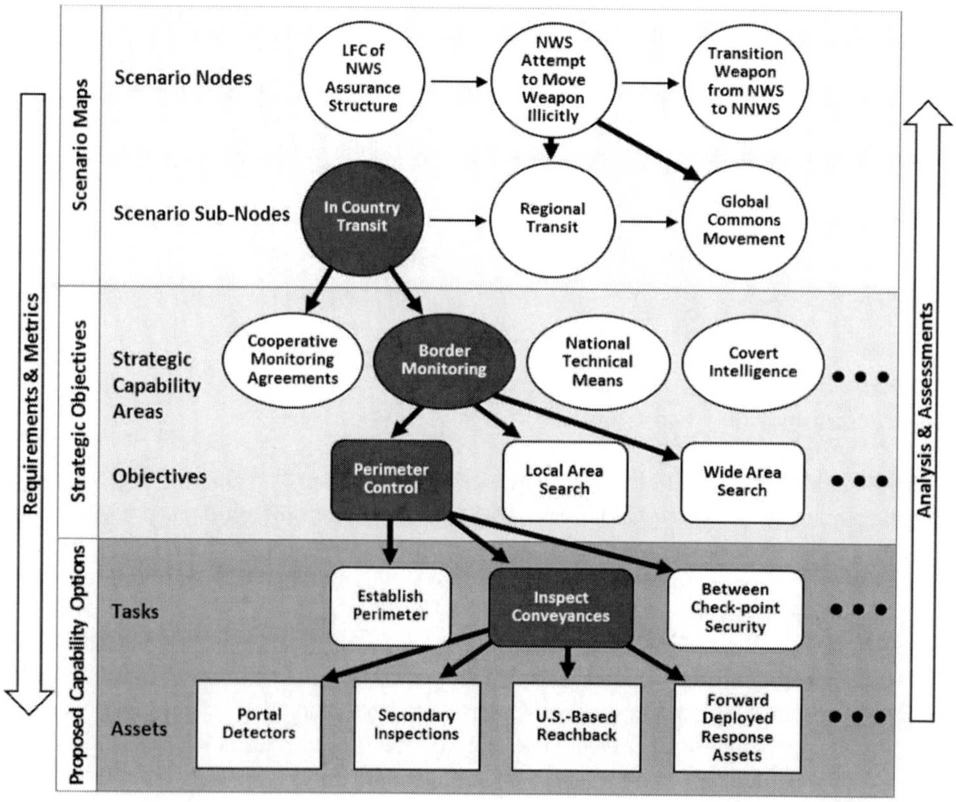

Figure A-3. Example Decomposition Map.

The derivation of a tractable set of metrics through this process is not a trivial task, nor should it be. Risk is often considered the top-level metric (with further discussion in the next section). Moving down the chain through strategic capability areas, functional objectives, tasks, and assets, metrics of increasing resolution and granularity are derived. At the asset level, the metrics center largely on performance specifications that are technology specific. These metrics are familiar in radiation detector assessments, but on their own, are only implicitly related to the overall goals of risk-reduction. A metrics derivation process such as this places each metric in the context of the layer above it, explicitly linking it to overall architecture performance.

A.7. Portfolio Decision Methodologies

The scenario framework, the bridging methodology, and the risk considerations proposed above lay out a systematic method for developing and assessing M&V solution architecture options. In order to render those results into investment roadmaps to develop capabilities, a decision framework must be established. While this section does not attempt to propose a decision framework, it does provide some considerations for doing so.

A.8. End-to-End Metrics

A significant challenge in adopting the approach that has been laid out above has been the articulation of an end-to-end metric that can be used to understand overall solution architecture performance against the problem space. Ultimately, it is that metric that matters most to decision makers. Risk provides a clear metric for endogenous trade-offs within the M&V problem space, allowing for the comparison of very different solution sets and examination of benefit between investments both within, and across, different components of the problem space itself. Utilizing risk as an end-to-end metric in the M&V problem space can also enable exogenous trades, as governments face economic challenges and must make tougher decisions about where to invest resources.

Utilization of risk as an end-to-end problem space metric comes with a set of inherent challenges as well. Common criticisms of formal risk assessment methodologies in decision processes include:

- **Conflating stochastic processes and adversary decisions** – Rigorous risk assessment methodologies rely on estimates of the probabilities of events of concern. In systems where outcomes are determined by truly random processes, this works quite well, especially when the system is well characterized. However, well characterized stochastic processes do not govern intelligent adversaries; instead, they make informed decisions. Although frequently used, probabilistic representations of adversary decisions are, for the most part, meaningless. However, characterization of uncertainty about adversary decisions in a probabilistic analysis can be beneficial, if carefully developed.
- **Focusing on absolute values rather than relative impacts and sensitivities** – The absolute values of risk are, in most formulations, arbitrary, as they are built upon the

assumptions and values of the analyst or decision maker for whom they are constructed. Additionally, the models upon which risk is calculated often cannot be truly validated. Therefore, the absolute values of risk are less important and somewhat meaningless, while the differences and relative comparisons can be more telling.

- **Inability to define "acceptable"** – A key component of making decisions in a risk-based framework is to define acceptable risks within the timeframe of the investment decision itself. If "acceptable" cannot be defined, the goal becomes to simply minimize risk, expending all resources, rather than achieve minimum risk using appropriate resources.
- **Examining risk trade-offs too narrowly** – One of the most common criticisms of implementations of risk-based approaches is the lack of definition around uncertainties. Especially in problems of high uncertainty, the error in risk calculations can be so large, that close trade-offs can be interpreted as essentially the same. Not having well defined uncertainties can allow false comparisons to drive decision processes. At the same time, ignoring the uncertainty (or error bars) in the analysis denies they opportunity to prioritize potential efforts to improve understanding.
- **Misuse of probability and statistics** – While it may seem like an elementary mistake, bad assumptions or interpretations, especially around dependence or independence, can lead to mathematical operations and inferences that may be numerically correct, yet meaningless—or worse, incorrectly calculated.

In the face of those challenges, it is certainly possible, although difficult, to articulate a risk metric that serves as an end-to-end metric within the M&V problem space for evaluating proposed solution architectures. It may not be necessary to always use the classical definitions of risk as the primary measure. Most risk studies use the arithmetic product of probabilities of occurrence and consequences as the standard definition of risk. However, the strictest interpretation of risk may not always be the most useful or accurate. Instead, it is possible to use measures that are proximate to risk when performing analyses. For example, other studies have proposed and formalized using assessments of the difficulty an adversary would face in executing a successful attack[41] to replace probability in the classical formulation of risk, when working with risk assessments of intelligent adversaries. Additionally, understanding the readiness level of an adversary and the expected time to attack can also serve as a useful risk proxy. Regardless of the exact risk metric or proxy used, risk (or some formulation of risk proxies) is likely to be a "necessary evil" as it allows for disparate approaches to be transacted together in cost-benefit trade-offs, and provides a consistent metric and method for analysis as new information and options become clear.

A.9. Balancing Risk

A key role of a portfolio decision methodology is to identify investments in capabilities that balance risks given the values of the decision maker. Investment risk does not only include cost and technical risk, but also institutional decision factors. Institutional decision factors are elements such as championship, mission, acceptance, and other factors that may prevent investments from being successful if not present. While not often considered in formal decision processes, they can have a large effect on the outcomes of decisions to acquire capabilities. Accounting for these factors in an M&V portfolio decision methodology is recommended.

A method for achieving a "balanced" capability portfolio can be desirable to some decision makers. For the purposes of this work, "balance" is defined as having an appropriate mix of capability investments to maximize risk reduction in the outcomes, while being prudent about the investment risks associated with each portfolio asset. As an analyst considers capability options, a balanced portfolio may be very different given the values and risk tolerance of the decision maker and the investing agency. For example, R&D focused organizations may have a set of values that is more risk tolerant, and a balanced investment portfolio may be skewed towards the long-term, high risk, high payoff projects. At the same time, an operational component may seek primarily commercial-off-the-shelf solutions and a balanced investment portfolio may be skewed towards incremental, low risk improvements. Formal and quantitative methods for assessing the risk tolerance and the values of decision makers exist and should be extended in formulating an M&V portfolio decision methodology.

APENDIX B. TERMS OF REFERENCE

THE UNDER SECRETARY OF DEFENSE
3010 DEFENSE PENTAGON
WASHINGTON, DC 20301-3010

ACQUISITION, TECHNOLOGY AND LOGISTICS

APR 2 6 2010

MEMORANDUM FOR CHAIRMAN, DEFENSE SCIENCE BOARD

SUBJECT: Terms of Reference-Defense Science Board (DSB) Task Force on the Assessment of Nuclear Treaty Monitoring and Verification Technologies

During the coming years, the United States is expected to engage in a series of treaty negotiations on nuclear weapons and nuclear forces. In addition, the rapid growth in nuclear power worldwide will likely stress the implementation practices of existing material control agreements, as well as poise more nations with the ability to acquire nuclear weapons of their own. Monitoring and verification measures are an integral part of all the existing, modified, or new agreements. Potential requirements for new or expanded monitoring and verification requirements place a renewed focus -after almost 2 decades of limited investment- on the adequacy of the Nation's technical tools to support monitoring and verification, both as part of the cooperative verification regimes of the treaties and through national intelligence.

You are to form a DSB Task Force to accomplish the following:

1. Research and summarize anticipated directions in nonproliferation and arms control agreements and the environments in which they might be implemented (for example, the level of transparency and cooperation that will be desired/required in post Cold

War arms control agreements, including treaties among nuclear states in addition to the United States and Russia);

2. Project the demands and challenges placed on existing agreements enforced by the International Atomic Energy Agency with the growth in nuclear power over the next 15 to 20 years, and assess the adequacy of current practices and resources to maintain confidence that inspected nations remain non-proliferators;
3. Assess current and programmed technical capabilities and gaps to support the anticipated monitoring and verification regimes;
4. Address what could be adapted from technical advances made for a number of other existing monitoring purposes, such as those developed for applications related to the following:
   a. Close-in and/or unmanned intelligence, surveillance, and reconnaissance systems to support conventional warfighting and counter-improvised explosive device operations;
   b. Stockpile stewardship;
   c. Nuclear forensics and attribution;
   d. Nuclear weapons effects; and
   e. Nuclear defense and interdiction programs.
5. Propose new initiatives, including identification of the technologies, research, development and testing program, and red/blue teaming requirements to fully vet new concepts; and
6. Perform a net assessment to understand the potential limitations and regrets associated with possible technical implementation paths.

The Task Force should recommend a comprehensive set of time-phased technical programs that could be conducted by the Department of Defense, the Department of Energy, the Intelligence Community, or a combination of these agencies, with consideration of what other agencies might also require (e.g., Department of State), contribute (e.g., Department of Homeland Security), or enable (e.g., Office of Science Technology and Policy).

The study will be co-sponsored by me as the Under Secretary of Defense for Acquisition, Technology and Logistics; the Administrator of the National Nuclear Security Administration (NNSA); and the Director of National Intelligence (DNI). Dr. Miriam John and Dr. Donald Kerr will serve as Chairpersons of the Task Force. Mr. Rhys Williams ofNNSA, Dr. Tom Hopkins from DATSD(NCB), and Mr. Michael Toomey ofODNI will serve as co-Executive Secretaries, and Major Michael Warner, USAF, will serve as the DSB Secretariat Representative.

The Task Force will operate in accordance with the provisions of P.L. 92-463, the "Federal Advisory Committee Act," and DoD Directive 5105.4, the "DoD Federal Advisory Committee Management Program." It is not anticipated that this Task Force will need to go into any "particular matters" within the meaning of title 18, U.S. Code, section 208, nor will it cause any member to be placed in *the* position of acting as a procurement official.

Ashton B. Carter

APPENDIX C. TASK FORCE MEMBERSHIP

Co-Chairs	
Dr. Miriam John	Independent Consultant
Hon. (Dr.) Donald Kerr	Independent Consultant
Executive Secretaries	
Dr. Tom Hopkins	Office of the Secretary of Defense
Mr. Michael Toomey	Office of the Director of National Intelligence
Dr. Rhys Williams	National Nuclear Security Administration
Members	
Dr. Joseph V. Braddock	Independent Consultant
Mr. John A. Lauder	Arete Associates
Dr. Anthony Pensa	Lincoln Laboratory
Hon. F. Whitten Peters	Williams & Connolly
Amb. (Dr.) Paul Robinson	Independent Consultant
Dr. Alexander C. Livanos	Northrop Grumman Corporation
Mr. Darryl Sargent	Draper Laboratory
Dr. James Tagnelia	Independent Consultant
Dr. Bruce Tarter	Lawrence Livermore National Laboratory
Mr. Lou Von Thaer	General Dynamics
Dr. Richard L. Wagner, Jr.	Independent Consultant
Dr. Joan Woodard	Independent Consultant
Senior Advisors	
Dr. Craig Fields	Independent Consultant
Hon. (Dr.) John S. Foster, Jr.	Independent Consultant
Hon. (Dr.) William Schneider Jr.	Independent Consultant
Government Advisors	
Mr. Richard Benson	Air Force
Mr. Bob Blum	Department of State
Mr. Doug Bruder	Defense Threat Reduction Agency
Mr. Don Cobb	Defense Threat Reduction Agency
Dr. Pat Falcone	White House, Office of Science & Technology Policy
Mr. Dick Gullickson	Defense Threat Reduction Agency
Mr. T.R. Koncher	Department of State
Mr. George Look	National Security Council
Mr. David O'Brien	Air Force
Dr. Joan Pierre	Defense Threat Reduction Agency
Mr. R.C. Porter	Defense Intelligence Agency
Dr. Caroline Purdy	Department of Homeland Security
Mr. Luis Salcedo	Navy
Mr. Kurt Seimon	National Nuclear Security Administration
Ms. Donna Smith	National Nuclear Security Administration
Lt. Col. John Smith, USAF	Office of the Secretary of Defense
Mr. Tom Troyano	Office of the Secretary of Defense

(Continued)

Mr. Larry Turnbull	UCIA
Laboratory Advisors	
Dr. Kory Budlong-Sylvester	Los Alamos National Laboratory
Dr. Mona Dreicer	Lawrence Livermore National Laboratory
Ms. Nancy Jo Nicholas	Los Alamos National Laboratory
Mr. Jason Reinhardt	Sandia National Laboratory
Dr. Sheila Vaidya	Lawrence Livermore National Laboratory
Mr. Bruce Walker	Sandia National Laboratory
DSB Secretariat	
Mr. Brian Hughes	Defense Science Board
Lt. Col. Michael Warner, USAF	Defense Science Board
CDR Doug Reinbold, USN	Defense Science Board
Contract Support	
Mr. Chris Grisafe	SAIC
Ms. Tammy-jean Beatty	SAIC

APPENDIX D. TASK FORCE MEETING SCHEDULE

Date	Briefing Title	Briefer	Affiliation
7/1/2010	DSB Intro Task Force Brief Arms Control and	Maj. Mike Warner	DSB
7/1/2010	Nonproliferation Priorities in the Administration	Mr. George Look	NSC
7/1/2010 7/1/2010 7/1/2010	New Arms Control Challenges: The Role of Intelligence in Verification and Monitoring Monitoring R&D Program International Monitoring System	Mr. John Lauder Mr. Dick Gullickson Mr. Don Phillips	Task Force Member DTRA Space and Missile Defense Command
7/1/2010 7/1/2010 7/2/2010	U.S. Atomic Energy Detection System Does Monitoring Matter? Intelligence Science Board	Dr. David O'Brien Dr. Joe Braddock Dr. Joan Woodard	AFTAC Task Force Member Task Force Member
7/2/2010	Limitations on UGT monitoring	Dr. Jay Zucca	LLNL
7/2/2010 7/22/2010	Space and Seismic Monitoring R&D programs Where the Materials Are in the world	Mr. Randy Bell Ms. Jamison Manternach	NNSA DOE
7/22/2010	DOE Global Threat Reduction Initiative	Mr. Ken Sheely	DOE
7/22/2010 7/22/2010	Technologies for Sustainability of Next Generation Russian Programs Program Brief: International Material Protection	Mr. Mike O'Brien Mr. John Gerrard	LLNL DOE
7/22/2010 7/22/2010 7/22/2010	IAEA Safeguards Cooperative Threat Programs Insights on Global Threat Reduction	Dr. Kory Budlong-Sylvester Amb. Bonnie Jenkins Dr. Joe Braddock	LANL DOS Task Force Member

Date	Briefing Title	Briefer	Affiliation
7/23/2010	Materials Protection	Ms. Laura Holgate	National Security Council
7/23/2010	Global Nuclear Lockdown	Mr. Tom Skillman	DoD
7/23/2010	Value of Monitoring and Verification	Hon. John P. Stenbit	Private Consultant
7/23/2010	Importance of Understanding NW Effects in Support of Verification	Dr. Joan Woodard	Task Force Member
8/24/2010	Role of TRMG and New START	Dr. Robert Blum and Mr. Ed Fortier	DOS
8/24/2010	CTBT NIE	Mr. Brian Lessenberry	NIC
8/24/2010	Nuclear Explosion Monitoring Challenges for the Next Decade	Dr. Larry Turnbull	JAEIC
8/24/2010	Test Monitoring Limitations	Mr. Don Linger	DTRA
8/24/2010	Russia's Stockpile	Ms. Marysusan Lynch	Office of Russian and European Analysis
8/24/2010	Rest of the World Stockpile and the Threat	Mr. John Galascione	WINPAC
8/24/2010	Importance of Understanding NW Effects in Support of Verification	Dr. Joan Woodard	Task Force Member
8/25/2010	Assessing Foreign Nuclear Programs	Mr. Bruce Held	Lab Advisor
8/25/2010	Nuclear Denial and Deception	Dr. Larry Gershwin	
8/25/2010	Nuclear Terrorism	Ms. Tara Swersie	NCTC
8/25/2010	Insights on Global Threat Reduction	Dr. Joe Braddock	Task Force Member
8/25/2010	Task Force Briefing v1	Dr. Mim John	Co-Chair
8/25/2010	Delivery Systems Chart	Dr. Mona Dreicer	Lab Advisor
9/13/2010	Future Issues for Safeguards	Amb Greg Schulte	DOS
9/13/2010	NK Nuclear Activity, CP Issues Worldwide and New START/NIE	Amb Joseph DeTrani and Mr. Bob Walpole	NCPC
9/13/2010	Open Skies	Mr. Scott Simmons; Mr. Dennis Connaughton; CDR Darin Liston; Mr. Don Spence & Mr. Michael Bett; Mr. Bob Harmon	DIA, DTRA, DOS, NGA
9/13/2010	Training for On-Site Inspections	Mr. Hunter Lutinski & Mr. Mark Beddoes	DTRA
9/13/2010	Next Generation Safe Guards	Mr. Steven LaMontagne	
9/13/2010	Task Force Briefing v2	Dr. Mim John	Co-Chair
9/14/2010	Monitoring & Verification for Arms Control (Password is: DSB)	Dr. Mona Dreicer	Lab Advisor
9/14/2010	START 3/Technology Development 1990-Present	Mr. John Dunn	NNSA
9/14/2010	U.S.-UK Technical Cooperation to Support Arms Reduction Initiatives	Ms. Michele Smith, NNSA	NNSA
9/14/2010	Nuclear Power	Dr. Kory Budlong-Sylvester	LANL
10/25/2010	Historical Perspectives on Proliferation and Perspectives on China	Mr. Brad Roberts	DASD, OSD/Policy

(Continued)

Date	Briefing Title	Briefer	Affiliation
10/25/2010	Sufficiency of R&D Activities Across Gov't	Dr. Pat Falcone	OSTP
10/25/2010	Data Processing for Wide Area Surveillance	Dr. Sheila Vaidya	LLNL
10/25/2010	Army Asymmetric Warfare Group	LTC Tom Newman, LTC David Wright	USA
10/25/2010	Technology Exploration	Dr. Joe Braddock	Member
10/26/2010	Hyperspectral Capability	Mr. Christopher Simi	Briefer
10/26/2010	Open Source Center Capabilities	Mr. Winston Dunleer, Mr. Darko Gerovac	Briefer
10/26/2010	Constant Hawk and Parallel Sensors	Mr. Steven Gotoff	DIA
10/26/2010	Task Force Briefing v3	Dr. Mim John	Co-Chair
1/6/2011	Technologies for Monitoring and Verification	Mr. Darryl Sargent	Draper Laboratory
1/6/2011	Wide Area Surveillance System Based Multi-INT Integration and Analysis for Situational Awareness and Knowledge Distillation	Dr. David Cremer	LANL
1/6/2011	Fast Fission Neutron Time Correlations from Fission for Non-Proliferation and Arms Control Applications	Dr. Leslie F. Nakae	LLNL
1/6/2011	Nonproliferation monitoring with distributed sensor networks	Dr. Simon Labov	LLNL
1/6/2011	The potential of small satellites for application to persistent monitoring and verification issues	Dr. Scot S. Olivier	LLNL
1/26/2011	Task Force Briefing v4	Dr. Mim John	Co-Chair
1/25/2011	Gamma Ray and Neutron Detection Methods for Nuclear Materials	Dr. Glenn Knoll	University of Michigan
1/25/2011	Domestic Nuclear Detection Office	Joel Rynes, Ph.D., PMP Brendan Plapp, Ph.D. Caroline Purdy, Ph.D.	DHS
1/25/2011	Radiation Detector Materials	Dr. Steve Payne	LLNL
1/25/2011	Radiation Detection at Sandia	Mr. Jim Lund	Sandia
1/25/2011	Laboratory Directed Research and Development: An Open Market for Ideas, Driven by Mission	Dan Holden	LANL
1/25/2011	Radiation Detection at NA-22	David Beach, Peter Vanier, Arden Dougan	NNSA
1/25/2011	DTRA Nuclear Detection	Dr. Peter Zielinski	DTRA
1/25/2011	Global Nuclear Architecture Report	DNDO	DHS
3/2/2011	Part 2. Framework end-to-end	Dr. Joan Woodard, Mr. Jason Reinhardt	Member/Lab Advisor
3/2/2011	Part 3. Sense and Assess – Rad detection	Dr. Bruce Tarter & Dr. Rhys Williams	Member/Exec Sec
3/2/2011	Part 3. Sense and Assess – Characterizing capabilities, "patterns of life"	Dr. Joe Braddock, Mr. John Lauder, Mr. Mike Toomey	Member/Exec Sec

Date	Briefing Title	Briefer	Affiliation
3/2/2011	Part 3. Sense and Assess – Persistent ISR	Mr. Lou Von Thaer, Dr. Tony Pensa, Dr. Sheila Vaidya	Member/Lab Advisor
3/3/2011	Part 3. Sense and Assess – Multi-INT analyses	Mr. Mike Toomey	Exec Sec
3/3/2011	Part 4. Access – Joint Experiments	Dr. Paul Robinson, Dr. Mona Dreicer	Member/Lab Advisor
3/3/2011	Part 4. Access – Open Skies	Dr. Jim Tegnelia, Mr. John Lauder	Member
3/3/2011	Part 4. Access – Materials control	Dr. Kory Budlong-Sylvester	Lab Advisor
3/3/2011	Part 4. Access – Dismantlement and Weapons Modernization Programs	Dr. Nancy Jo Nicholas, Mr. Bob Huelskamp, Dr. Rich Wagner	Member/Lab Advisor
3/3/2011	Part 4. Access -Non-cooperative environments	Mr. Darryl Sargent	Member
3/3/2011	Part 5. Iterate and Adapt	Dr. Rich Wagner, Dr. Jim Tegnelia, Dr. Joe Braddock	Member
3/3/2011	Adapt & Iterate Paper	Dr. Joe Braddock	Member
3/3/2011	DOTMLPF-Theater Nuclear Forces Paper	Dr. Joe Braddock	Member
3/3/2011	Non-Traditional Signatures	Dr. Joe Braddock	Member
3/3/2011	Part 6. Leadership	Mr. Whit Peters, Mr. John Lauder, Dr. Mim John, Hon. Don Kerr	Member/Co-Chair
3/3/2011	Detecting Nuclear Threat	Dr. Michael A. Kuliasha	DTRA
3/3/2011	STASIS	Dr. Jasper Lupo	Applied Research Associates
3/3/2011	Task Force Briefing v5	Dr. Mim John	
3/3/2011	Open Source Center	Mr. Mike Toomey	
5/3/2011	General Story Line and Part 1. Introduction	Dr. Mim John, Hon. Don Kerr	
5/3/2011	Part 4. Sense and Assess – Rad detection	Dr. Rhys Williams, Dr. Don Cobb, Bruce Tarter	
5/3/2011	Part 2. Framework end-to-end	Dr. Joan Woodard & Mr. Jason Reinhardt	
5/3/2011	Part 3. Monitoring Architecture: Access-Sense-Assess-Iterate	Mr. Lou Von Thaer	
5/3/2011	Part 4. Sense and Assess: NTM: its role and its limitations	Mr. John Lauder and Dr. Don Kerr	
5/3/2011	Part 4. Sense and Assess: Signs of activities Associated with Having a Viable Capability	Mr. Mike Toomey, Dr. Rhys Williams & Dr. Joe Braddock	
5/3/2011	Part 4. Sense and Assess: Initial filter: "Patterns of life" analyses	Mr. John Lauder & Mr. Mike Toomey	
5/3/2011	Part 4. Sense and Assess: DOTMLPF	Dr. Joe Braddock	
5/4/2011	Part 4. Sense and Assess: Next tier of observation: Persistent ISR	Mr. Lou Von Thaer & Dr. Tony Pensa	

(Continued)

Date	Briefing Title	Briefer	Affiliation
5/4/2011	Part 4. Sense and Assess: Developments for counter-IED problem (COIC)	Group discussion re what we should say about this	
5/4/2011	Part 4. Sense and Assess: Large data set processing	Dr. Sheila Vaidya & Dr. Tony Pensa	
5/4/2011	Part 4. Sense and Assess: "nth-country/group" analytical capability evolved for today's problem space	Dr. Mona Driecer	
5/4/2011	Part 4. Sense and Assess: Concept for the CWMD-COP	Dr. Jim Tegnelia	
5/4/2011	Part 5. Access: Whether facing cooperative or non-cooperative environments, we need more options to improve access	Dr. Paul Robinson, Dr. Mona Dreicer & Kory Budlong-Sylvester	
5/4/2011	Part 6. Experiment, Iterate and Adapt	Dr. Rich Wagner, Dr. Jim Tegnelia & Dr. Joe Braddock	
5/4/2011	Clarifying terms: Monitoring and Verification	Paul Robinson	
5/4/2011	the "Building Transparency" Recommendation	Paul Robinson & Mona Dreicer	
5/4/2011 5/4/2011 5/4/2011	Open Source Reporting and Analysis Testing Capability Chart Open Source Center Text	Mike Toomey Joe Braddock Mike Toomey	

APPENDIX E. ACRONYMS USED IN THIS REPORT

AF	Air Force
ASD(NCB)	Office of the Assistant Secretary of Defense for Nuclear, Chemical and Biological Defense Programs
ASW	Anti-Submarine Warfare
AT&L	Acquisition, Technology, and Logistics
AVC	Bureau of Arms Control, Verification and Compliance
CD	Compact Disc
CIS	Commonwealth of Independent States
COCOMs	combatant commands
COIC	Counter IED Operations Integration Center
CONOPs	Concept of Operations
Counter-IED	Countering Improvised Explosive Devices
CP	Counter proliferation
CPP	Counterproliferation Program
CTBT	Comprehensive Nuclear-Test-Ban Treaty
CTR	Cooperative Threat Reduction
DARPA	Defense Advanced Research Projects Agency
DC	Dual Capable

DCA	Dual-capable aircraft
DHS	Department of Homeland Security
DNDO	Domestic Nuclear Detection Office
DNA	Defense Nuclear Agency
DNI	Director of National Intelligence
DOD	Department of Defense
DOE	Department of Energy
DOS	Department of State
DTRA	Defense Threat Reduction Agency
EUCOM	United States European Command
FFRDCs	Federally Funded Research and Development Centers
GCAS	Global Counter Weapons of Mass Destruction Awareness System
GEOINT	Geospatial Intelligence
GLCM	Ground launched cruise missile
GNDA	Global Nuclear Detection Architecture
HEU	Highly-Enriched Uranium
HUMINT	Human Intelligence
IAEA	International Atomic Energy Agency (IAEA)
IC	Intelligence Community
IED	Improvised Explosive Device
INFCIRC	International Atomic Energy Agency Information Circular
INT	Intelligence
IWG	Interagency Working Group
ISAB	International Security Advisory Board (Department of State)
ISR	Intelligence, Surveillance, and Reconnaissance
IT	Information Technology
JVE	Joint Verification Experiment
M&V	Monitoring and Verification
MASINT	Measurement and Signature Intelligence
MOU	Memorandum of Understanding
NATO	North Atlantic Treaty Organization
NCPC	National Counterproliferation Center
NGA	National Geospatial-Intelligence Agency
NGO	Non-governmental Organization
NNSA	National Nuclear Security Agency
NNSA/NA-20	NNSA Defense Nuclear Non-Proliferation
NNSS	Nevada National Security Site
NNWS	Non-Nuclear Weapons States
NPT	Non-Proliferation Treaty or Treaty on the Non-Proliferation of Nuclear Weapons
NSS	National Security Staff
OSC	Open Source Center
OSINT	Open Source Intelligence
OSTP	Office of Science & Technology Policy
P-5	Five Permanent Members of the United Nations Security Council

POET	Phase One Engineering Team
Pu	Plutonium
R&D	Research and Development
RDT&E	Research, Development, Test & Evaluation
RSTA	Reconnaissance, surveillance, and target acquisition
S&T	Science and Technology
SDIO	Strategic Defense Initiative Organization
HAPE	Supreme Headquarters Allied Powers Europe
SIGINT	Signals Intelligence
SNM	Special Nuclear Material
SSD	Safe, Secure Dismantlement
START	Strategic Arms Reduction Treaty
TMM	Treaty Monitoring Manager
TNF	Theater Nuclear Forces
TOR	Terms of Reference
TOSI	Technical On-Site Inspection
TTPs	Tactics, techniques and procedures
U.S.	United States
UAV	Unmanned Aerial Vehicle
UGS	Unattended Ground Sensor
USAEDS	U.S. Atomic Energy Detection System
USG	United States Government
VCJCS	Vice Chairman of the Joint Chiefs of Staff
WMD	Weapons of Mass Destruction
WP	Soviet/Warsaw Pact
WSSX	Warhead Safety and Security Exchange
WUNM	Weapons Usable Nuclear Material

End Notes

[1] See Section 5 for a more complete discussion on the Mission Manager role.

[2] See next section "Address the Problem Whole."

[3] The responsibilities of the Treaty Monitoring Manger are discussed more fully in the context of proliferation monitoring in Section 5.

[4] An excellent recent example of this is the "Nuclear Defense Research and Development Roadmap. Fiscal Years 2013-2017" published by the National Science and Technology Council, April 2012.

[5] Nikolai Sokov, Russia's New National Security Concept: The Nuclear Angle, Center for Nonproliferation Studies Report, January 2000.

[6] Statement by Sergey Ivanov, available at http://www.mil.ru/articles/article3667.shtml

[7] Vladimir Putin, "Zakluychitelnoe Slovo na Soveshchanii s Rukovodyashim Sostovom Vooruzhennykh Sil Rossii", October 2, 2003 (available at htpp://www.president.kremlin.ru/text/appears/2003/10/53277.shtml).

[8] Yuriy Golotuyk, "I v Vozdukhe Tozhe Problemy" Vremya novostey, February 19, 2001.

[9] "Strategicheskaya Komandno-Shtabnaya Treniroivka VS Rossii", Nezavisimaya gazeta, February 17, 2001.

[10] These kinds of connections and synergies in monitoring are developed in more detail for the case of monitoring of Theater Nuclear Forces (TNF) and dual capable (DC) systems later in the report [Section 6, section 6.2.5].

[11] These approaches are discussed in more detail in Section 3, section 3.4, and Section 5.

[12] The national testing capability is discussed in some detail in Section 6, sections 6.2 and 6.3.

[13] The NPT is a landmark international treaty whose objective is to prevent the spread of nuclear weapons and weapons technology, to promote cooperation in the peaceful uses of nuclear energy and to further the goal of achieving nuclear disarmament and general and complete disarmament. The Treaty represents the only binding

commitment in a multilateral treaty to the goal of disarmament by the nuclear-weapon States. Opened for signature in 1968, the Treaty entered into force in 1970. On 11 May 1995, the Treaty was extended indefinitely. A total of 190 parties have joined the Treaty, including the five declared (at that time) nuclear-weapon States. More countries have ratified the NPT than any other arms limitation and disarmament agreement, a testament to the Treaty's significance. (Ref.: http://www.un.org/disarmament/WMD/Nuclear/NPT.shtml)

[14] Then Senator Sam Nunn and Senator Richard Lugar initiated the program at the request of Secretary William Perry and Assistant Secretary of Defense Ashton Carter. Both Nunn and Lugar have remained active participants in the program over the years, visiting CTR sites in the former Soviet Union and guiding and encouraging aggressive conduct of CTR activities. They have also been instrumental in solving intergovernmental problems that otherwise would have delayed progress on the program.

[15] In addition, the CTR program assisted the Russians in making a best effort to meet its obligations under the Chemical Weapons Convention. The United States assisted the Russians in building chemical demilitarization facilities that destroyed thousands of chemical weapons. It has been expanded in recent years to secure biological agents of concern, as well as expand disease monitoring capabilities in selected international regions.

[16] The IAEA does publish the plutonium holdings of nine countries, the five declared weapons states as well as Japan, Germany, Belgium and Switzerland. This information is provided on a voluntary basis by the participants consistent with the "Guidelines for the Management of Plutonium" (INFCIRC/549). Such declarations might be a useful starting point upon which to build a more comprehensive transparency regime.

[17] The 25th Anniversary of the Joint Verification Experiment was commemorated jointly between the U.S. and Russia in the early fall of 2013 at the site of the experiment.

[18] These issues have also been recognized by the State Department's International Security Advisory Board (ISAB), and near term technical steps for trust building with Russia proposed. See ISAB report "Verification Measures - Near Term Technical Steps," 2012; http://www.state.gov/t/avc/isab/200465.htm

[19] See Sections 3 and 5 for further discussion on this topic.

[20] See Section 5 for a more complete discussion on the Mission Manager role.

[21] See Sections 4 and 6 for a more complete discussion on the Blue and Red teaming.

[22] Section 123 of the *United States Atomic Energy Act of 1954*, titled "Cooperation With Other Nations", establishes an agreement for cooperation as a prerequisite for nuclear deals between the US and any other nation. Such an agreement is called a 123 Agreement. To date, the U.S. has entered into roughly twenty-five 123 Agreements with various countries.

[23] The potential for OSINT (open source intelligence) as a focusing step is also promising and should be explored. See Section 5, Section 5.5.4 for a discussion.

[24] The responsibilities of the Treaty Monitoring Manger are discussed more fully in the context of proliferation monitoring in Section 5.

[25] This task force built upon previous work for OSD/NCB/NM by the DOE/NNSA laboratories: "*Analysis of Capability Options for the DoD Countering Nuclear Threats Mission: NNSA Tri-lab Phase 1 Summary,*" Sandia National Laboratories report SAND 2011-7985, November 2011.

[26] Wyss, Hinton, et al, Risk-Based Cost Benefit Analysis for Security Assessment and Investment Prioritization, Sandia National Laboratories, 2011For example:

[27] NNSA, "Special Nuclear Materials Detection Program: Radiation Sensors and Sources Roadmap," NA22-OPD-01-2010 (recommended as one of the most comprehensive)

[28] Congressional Research Service, "Detection of Nuclear Weapons and Materials:Science, Technologies, Obervations," R40154, June 4, 2010

[29] JASONS, "Concealed Nuclear Weapons," JSR-03-130(2003); "Active Interrogation," JSR-09-2-2 (2009); "Lifetime Extension Program (LEP) Executive Summary," JSR-09-334E (2009)

[30] "Nuclear Forensics: A Capability at Risk," A. Carnesale (committee chair), National Research Council, 2010.

[31] The airborne component of the system was deployed and proved invaluable in the aftermath of the Fukishima nuclear reactor disaster.

[32] DARPA Red Balloon Challenge (http://archive.darpa.mil/networkchallenge/)

[33] M. Fisher, "Google Maps Reveals Exact Site of North Korea's Nuclear Test, Plus Nearby Test Facility and Gulag," Washington Post, February 12, 2013

[34] C. Hansell, Cristina; William C. Potter; "Engaging China and Russia on Nuclear Disarmament," Monterey Institute of International Studies, April 2009

[35] W. Wan, "Georgetown Students Shed Light on China's Tunnel System for Nuclear Weapons," *Washington Post,* November 29, 2011

[36] An excellent recent example of this is the "Nuclear Defense Research and Development Roadmap. Fiscal Years 2013-2017," National Science and Technology Council, April 2012.

[37] SORAK was a South Korean experimentation concept developed in the late 1970's and early 1980's based on Shock Wave. It was sponsored by Commander-in-Chief, U.S. Pacific Command (CINCPAC) and supported

by DARPA, as part of the joint United States-South Korean exercise Ulchi Focus Lens. The concept was intended to create an all source testing capability to observe operations.

[38] Pershing II and GLCM operations were driven by the need for survivability in moving units out of peacetime garrison in crisis and in move-and-hide tactics in the field in crisis and war. These operations were developed iteratively using red-teamed test-bed exercises coupled with simulations, both in CONUS and in Europe.

[39] Dunn, Lewis, *Global Shocks and Surprise: Shaping the 2030 Nuclear Future*, presented to the Joint Sandia Laboratories – University of California "Pathways to Alternative Nuclear Futures Workshop," September 2009; http://sandia.gov/nuclear_pathways/. In this paper, Dunn referred to scenarios as "shock- free surprises" that are distinct from truly "shocking surprises". Shocking surprises refer to events that may radically alter priorities of a decision maker, and that an analyst may not foresee as feasible. An M&V analytical methodology should not be exclusive of "shocking surprises."

[40] Danzig, Richard, *Catastrophic Bioterrorism – What Is To Be Done?*, Center for Technology and National Security Policy, August 2003. In this paper, Danzig stated that individual efforts tend to be unrelated to any overarching strategy, and measures of effectiveness are difficult to formulate. Danzig also observed that relevant tools and capabilities were not viewed as alternatives, or as complements, but rather as individual programs, each operating independently with resource allocations being made more in accord with bureaucratic position and power, rather than in response to the problem.

[41] Wyss, Hinton, et al, Risk-Based Cost Benefit Analysis for Security Assessment and Investment Prioritization, Sandia National Laboratories, 2011.

In: Nuclear Security
Editor: Charlotte R. Carey

ISBN: 978-1-63117-643-2

Chapter 2

Nuclear Nonproliferation: IAEA Has Made Progress in Implementing Critical Programs but Continues to Face Challenges*

United States Government Accountability Office

Why GAO Did This Study

IAEA plays a crucial role in supporting U.S. nuclear nonproliferation goals through its safeguards and nuclear security programs. The Department of State (State) coordinates the United States' financial and policy relationship with IAEA. IAEA's safeguards program is designed to detect and deter the diversion of nuclear material for non-peaceful purposes, while the agency's nuclear security program assists countries in improving the physical protection of their nuclear material and facilities. IAEA plans to create an international fuel bank to guarantee the supply of fuel for civilian nuclear power programs. GAO was asked to examine (1) any challenges that IAEA faces in carrying out its safeguards program, (2) any limitations regarding the nuclear security program, and (3) the status of IAEA's planned nuclear fuel bank. GAO reviewed relevant documents and interviewed officials from IAEA and U.S. federal agencies, 15 nuclear nonproliferation experts, and representatives from 16 countries.

What GAO Recommends

GAO recommends, among other things, that State work with IAEA to (1) clearly define and communicate how IAEA will implement the state-level concept, (2) evaluate the nuclear security program's long-term resource needs, and (3) prepare a plan for the long-term operation and funding of IAEA's fuel bank. State agreed with several of the recommendations and disagreed with two, including the one on the nuclear security program's long-term

* This is an edited, reformatted and augmented version of the United States Government Accountability Office publication, GAO-13-139, dated May 2013.

resource needs. GAO continues to believe that implementing all of these recommendations would enhance the efficiency and effectiveness of IAEA's programs.

WHAT GAO FOUND

The International Atomic Energy Agency (IAEA) has taken several steps to strengthen its safeguards program since GAO's 2005 report, including increasing the number of countries that are subject to a broader range of safeguards measures, upgrading its analytical laboratories, and producing its first long-term strategic plan. However, the agency faces two critical challenges in further improving the effectiveness and efficiency of the safeguards program. First, several countries have raised concerns about IAEA's plans to expand implementation of what it calls the "state-level concept" to all countries with safeguards agreements by 2014. The state-level concept is an approach in which the agency considers a broad range of information about a country's nuclear capabilities and tailors its safeguards activities in each country accordingly. IAEA officials told GAO that broader implementation of this approach would allow the agency to better allocate resources by reducing safeguards activities where there is no indication of undeclared nuclear activities and to focus its efforts on any issues of safeguards concern. However, IAEA has not clearly defined and communicated how it will implement the state-level concept. As a result, several countries are concerned that the state-level concept may be applied in a subjective, potentially discriminatory manner or that it could allow IAEA to be too intrusive into their civilian nuclear operations. Second, the agency has not quantified the resources it may need to fully implement the state-level concept— making it difficult to determine its long-term costs and benefits.

IAEA has continued to support countries' efforts to improve the security of their nuclear material and facilities, but three key issues limit the agency's ability to ensure that its nuclear security resources are used efficiently and effectively. First, IAEA's nuclear security program relies heavily on extra-budgetary contributions from donor countries, which makes it difficult to plan and implement projects, in part because these funds vary from year to year. Second, IAEA has not conducted a needs-based assessment of the resources required beyond its 2-year budget cycle, which also hinders its ability to ensure that resources are directed to the greatest security needs. Third, the extent to which IAEA is meeting its nuclear security goals is unclear because IAEA does not systematically report on the results of measures used to assess the performance of the agency's nuclear security program.

IAEA is making progress in establishing an international nuclear fuel bank by 2014 that is intended to provide eligible countries with a guaranteed supply of low enriched uranium (LEU) for civilian nuclear power programs in the event of a supply disruption. IAEA's fuel bank is not intended to provide a routine supply of LEU. However, several experts and foreign officials told GAO that the fuel bank's value is uncertain, causing some foreign officials to call it "a solution looking for a problem." For example, the fuel bank may never be used, in part because there are already several stable suppliers on the international nuclear fuel market. In addition, IAEA does not have a plan in place for the long-term operation and funding of the bank, although agency officials told GAO they intend to complete such a plan in 2013. Furthermore, the IAEA bank is one of several guaranteed, multilateral fuel supply options—including banks established by the United States and Russia—for countries seeking an assured supply of nuclear fuel.

ABBREVIATIONS

DOE	Department of Energy
HEU	highly enriched uranium
IAEA	International Atomic Energy Agency
IPPAS	International Physical Protection Advisory Service
LEU	low enriched uranium
NNSA	National Nuclear Security Administration
NPT	Treaty on the Nonproliferation of Nuclear Weapons
USEC	United States Enrichment Corporation
U.S. Mission	U.S. Mission to the United Nations System Organizations in Vienna

May 16, 2013

The Honorable Thomas R. Carper
Chairman
Committee on Homeland Security and Governmental Affairs
United States Senate

The Honorable Ron Johnson
Ranking Member
Subcommittee on Financial and Contracting Oversight
Committee on Homeland Security and Governmental Affairs
United States Senate

For more than half a century, the International Atomic Energy Agency (IAEA) has played a crucial role in supporting the nuclear nonproliferation goals of the United States and the international community through its safeguards and nuclear security programs.[1] IAEA's safeguards role—to detect and deter the diversion of nuclear material for non-peaceful purposes—is particularly relevant in light of major challenges that IAEA continues to face with Iran, Syria, and North Korea. IAEA has found these three countries to be in non-compliance with their obligations under their safeguards agreements with the agency, which are pursuant to the Treaty on the Nonproliferation of Nuclear Weapons (NPT). As a result, IAEA cannot provide assurance that there are no undeclared nuclear activities in these three countries. The NPT, which came into force in 1970, requires non-nuclear weapon states that are party to the treaty—countries that had not manufactured and detonated a nuclear device before January 1, 1967—to agree not to acquire nuclear weapons.[2] The NPT also requires these countries to subject all nuclear material used in peaceful activities to IAEA safeguards.[3] As of April 2013, 184 non-nuclear weapon states and 5 nuclear weapon states—China, France, Russia, the United Kingdom, and the United States—were parties to the NPT.[4]

In implementing the safeguards agreements required by Article III of the NPT, IAEA carries out several important safeguards activities to deter the proliferation of nuclear weapons and to ensure that nuclear material and facilities are used exclusively for peaceful purposes. As of December 2011, IAEA reported that 1,209 nuclear facilities and other locations outside such facilities containing significant quantities of nuclear material were subject to safeguards activities.[5] To carry out its safeguards activities, IAEA inspectors and analysts collaborate to verify that the quantities of nuclear material that non-nuclear weapon states have formally declared to the agency are correct and complete and have not been diverted for the manufacture of nuclear weapons, other nuclear explosive devices, or for unknown purposes. Most countries have negotiated an agreement with IAEA known as a comprehensive safeguards agreement, which covers all of their civilian nuclear activities and serves as the basis for the agency's safeguards activities. Specifically, IAEA's safeguards activities include on-site inspections, environmental sampling,[6] remote monitoring, analysis of commercial satellite imagery, and analysis of publicly available (open source) documents. In addition, the agency installs containment and surveillance measures, such as video cameras, to detect the movement of nuclear material and any tampering with agency equipment; as of November 2012, IAEA had installed 27,000 tamper-proof seals and about 1,200 cameras at nuclear facilities around the world. Pursuant to their safeguards agreements with IAEA, individual countries are obligated to implement specific safeguards activities, which vary from country to country based on the amount and type of nuclear material a country possesses and the type of nuclear facilities and fuel cycle activities a country operates. However, IAEA relies on the cooperation of individual countries for certain safeguards activities that are not legal obligations under safeguards agreements, such as granting IAEA permission to conduct remote monitoring, which may be domestically sensitive. On an annual basis, IAEA draws a conclusion for each country about whether a country is compliant with its safeguards agreement. Specifically, IAEA draws a conclusion for a given year regarding whether a country's declared nuclear material remained in peaceful activities and, where applicable, that there are no indications of undeclared nuclear material or activities.

In addition to its safeguards activities, IAEA's nuclear security work also plays an important role internationally. IAEA's nuclear security program— which assists countries in enhancing the physical protection, control, and accounting of their nuclear and radiological material and nuclear facilities—has assumed a higher profile since the September 11, 2001, terrorist attacks on the United States. Whereas IAEA's nuclear security program was previously housed within the Department of Safeguards, in 2002 IAEA created its Office of Nuclear Security within the Department of Nuclear Safety and Security. Since then, IAEA's role in nuclear security has grown, in concert with increased international efforts to strengthen nuclear security. For example, in March 2012, world leaders from more than 50 countries and international organizations met in Seoul, South Korea, for the second Nuclear Security Summit, where they renewed their political commitment to work toward strengthening nuclear security. To assist countries in strengthening nuclear security, IAEA offers, among other things, a variety of training courses that draw upon international obligations, recommendations, guidelines, and best practices.

To complement the agency's efforts in its safeguards and nuclear security programs, IAEA's Board of Governors—which provides overall policy direction and oversight—approved the creation of an international nuclear fuel bank in December 2010. This fuel bank is intended to provide a guaranteed supply of low enriched uranium (LEU) for civilian nuclear power in the event of a supply disruption; it is not intended to provide a routine supply.[7] As such, the establishment of an IAEA-owned and operated nuclear fuel bank is also intended to offer countries an alternative to developing their own indigenous fuel supplies, as the use of sensitive nuclear fuel enrichment technologies to develop fuel could also be used for a clandestine nuclear weapons program. Kazakhstan has volunteered to serve as the host state of the fuel bank.

In 2011, IAEA reported a regular budget of $433.1 million, of which the United States contributed $111.1 million, making it the largest contributor.[8] IAEA also reported receiving $317.3 million in 2011 in extra-budgetary cash contributions that a number of countries provided on a voluntary basis, in addition to their contributions to the agency's regular budget. According to the Department of State (State), the United States contributed $78 million in extra-budgetary cash contributions to the agency's various programs that year. The United States has a strong interest in IAEA's safeguards and nuclear security activities, as well as its implementation of the international nuclear fuel bank. We last reviewed IAEA's safeguards and nuclear security programs in 2005 and found a number of issues that limited the effectiveness and efficiency of the agency's efforts.[9] We recommended, among other things, that State— which coordinates the United States' financial and policy relationship with IAEA—work with IAEA to consider (1) eliminating or reducing the number of agreements that limit IAEA's authority to implement strengthened safeguards measures in countries with small quantities of nuclear material, (2) developing a systematic process that forecasts safeguards budgetary requirements and resources for the long-term, and (3) establishing better measures to evaluate the effectiveness of nuclear security activities. State agreed with these recommendations and has worked with IAEA to implement some of them. In this context, you asked us to update information on the IAEA safeguards and nuclear security programs, as well as provide information regarding the status of the international nuclear fuel bank program. Accordingly, our objectives were to examine (1) changes that have occurred since 2005 and challenges, if any, that IAEA faces in carrying out its safeguards program; (2) changes that have occurred since 2005 and limitations, if any, regarding IAEA's nuclear security program; and (3) the status of IAEA's planned nuclear fuel bank.

To address these objectives, we reviewed documents from IAEA, including the agency's biennial program and budget documents, annual reports, and long- and medium-term strategic plans. We also interviewed IAEA officials in Vienna, Austria, who are knowledgeable about the agency's safeguards, nuclear security, and fuel bank programs. In addition, we collected documents from and interviewed officials from State, the Department of Energy's National Nuclear Security Administration (NNSA),[10] and national laboratories. We also interviewed officials from a nonprobability sample of 16 countries—Brazil, Canada, Chile, China, France, Germany, India, Israel, Japan, Pakistan, Russia, South Korea, Turkey, the United Arab Emirates, the United Kingdom, and the United States—to obtain their views on all three programs.[11] We selected this diverse range of countries based on a number of factors, including, among other things, their civilian nuclear power capabilities, geographic location, and NPT status as nuclear weapon or non-nuclear weapon states. We also interviewed a nonprobability sample of 15 nuclear nonproliferation experts, including 4 former senior IAEA officials, for their knowledge of IAEA's safeguards, nuclear security, and/or fuel bank programs. We selected these experts based on several factors, including their recent prior work at IAEA, their relevant work at academic institutions or nongovernmental organizations, or referrals from other nuclear nonproliferation experts.[12] Throughout this report, we distinguish between statements made by the U.S. Mission to the United Nations System Organizations in Vienna (U.S. Mission) and the 15 other countries (foreign missions), and we use the term "some" when referring to two or three experts or countries and "several" when referring to four or more experts or countries. To assess the reliability of budget data related to IAEA's safeguards, nuclear security, and fuel bank programs, we compared data from different published sources, including the *Federal Register*, and we met with IAEA and U.S. officials to discuss these data in detail. Based on this work, we determined that data provided by IAEA and State to be sufficiently reliable for presenting funding information for the safeguards, nuclear security, and fuel bank programs. Appendix I provides a more detailed discussion of our objectives, scope, and methodology.

We conducted this performance audit from January 2012 to April 2013 in accordance with generally accepted government auditing standards. Those standards require that we plan and perform the audit to obtain sufficient, appropriate evidence to provide a reasonable basis for our findings and conclusions based on our audit objectives. We believe that the evidence obtained provides a reasonable basis for our findings and conclusions based on our audit objectives.

BACKGROUND

This section outlines IAEA's structure, budget, and the safeguards, nuclear security, and fuel bank programs.

IAEA Structure

IAEA's governing bodies include the General Conference, composed of representatives of 159 countries that contribute to IAEA's budget (member countries)[13] and the 35-member Board of Governors, of which the United States is a de facto permanent member. A Secretariat, headed by the Director General, is responsible for implementing the policies and programs of the General Conference and Board of Governors. U.S. policy regarding the agency is developed by an interagency process chaired by State. The U.S. Mission works closely with State in Washington, D.C., to promote the effective functioning of the agency, including management reform.

IAEA Budget

IAEA funds its programs primarily through (1) its regular budget, for which all member countries are assessed,[14] and (2) extra-budgetary cash contributions. In 2011, IAEA reported that its regular budget was $433.1 million, of which the safeguards program budget comprised $166.1 million and the nuclear security program comprised $5.3 million. The agency also receives extra-budgetary cash contributions from certain member countries, particularly the United States, and other donors on a voluntary basis, in addition to the agency's regular budget and to meet critical needs.[15] For example, in 2011, IAEA reported that the safeguards program received $39.3 million, and the nuclear security program received $25 million, in extra-budgetary cash contributions. Of these amounts, the United States made extra-budgetary cash contributions of $24.8 million and $13.4 million to these programs, respectively. (See table 1.) The agency has not used any regular budget funding for the fuel bank program but has received $159.4 million in extra-budgetary cash contributions and pledges toward the establishment of the fuel bank, of which the U.S. government contributed nearly $50 million in 2008, the largest single-country donation to the fuel bank.

IAEA has operated under a "zero-real-growth" budget policy environment for over 27 years,[16] which has generally caused IAEA's programs to experience minimal budgetary growth from year to year and to seek efficiencies on an ongoing basis. For example, IAEA reported in 2011 that inspectors spent fewer days in the field for certain safeguards activities, and the frequency of inspections was reduced, compared with the prior year. According to IAEA officials, those cost savings were offset by other measures to optimize safeguards activities, such as through remote monitoring or unannounced, random inspections.

Safeguards Program

IAEA derives its authority to establish and administer safeguards from its statute, the NPT and regional nonproliferation treaties, bilateral commitments between countries, and project agreements with countries. Article III of the NPT binds each of the treaty's 184 non-nuclear weapon state parties to conclude an agreement with IAEA that applies safeguards to all source and special fissionable material in all peaceful nuclear activities within the state's territory, under its jurisdiction, or carried out anywhere under its control.[17]

IAEA's safeguards objectives, as traditionally applied, are to account for the amount of special fissionable material necessary to produce a nuclear weapon, and the time it might take a country to divert this material from peaceful use and produce a nuclear weapon. IAEA attempts to meet these objectives by using a set of activities by which it seeks to verify that nuclear material subject to safeguards is not diverted to nuclear weapons, other proscribed purposes, or for unknown purposes. For example, IAEA inspectors visit a facility at certain intervals to ensure that any diversion of nuclear material is detected before a country using that material has had time to produce a nuclear weapon. IAEA also uses "material accounting" measures to verify quantities of nuclear material declared to the agency and any changes in the quantities over time.

The safeguards program has evolved over the past two decades as IAEA has made several efforts to strengthen its effectiveness and efficiency. Starting in the early 1990s, in response to the 1991 discovery of a clandestine nuclear weapons program in Iraq, IAEA made a radical departure from its past practice of generally verifying only the peaceful use of a country's declared nuclear material at declared facilities; at that time, IAEA expanded its safeguards efforts to detect potentially undeclared nuclear activities as well. Specifically, through its Department of Safeguards, the agency began exercising its existing authority under safeguards agreements with individual countries to obtain additional information about their nuclear and nuclear-related activities. However, IAEA recognized that these additional measures were not adequate. As a result, in 1997 the Board of Governors approved what it called the "Model Additional Protocol" (Additional Protocol)—which, when ratified or otherwise brought into force by a country, requires that country to provide the agency with a broader range of information on its nuclear and nuclear-related activities. It also gives the agency's inspectors access to an expanded range of declared activities and locations, including buildings at nuclear sites, and locations where undeclared activities may be suspected. A country agrees separately to an Additional Protocol, which supplements the country's existing safeguards agreement with IAEA.

At the time of our 2005 report, IAEA's safeguards program was in the process of strengthening its capability to independently evaluate all aspects of a country's nuclear activities through broader inspector access, collecting and analyzing environmental and nuclear material samples to verify that countries' nuclear material declarations are correct and complete, and analyzing open source information.[18] In addition, we reported that IAEA was trying to improve the efficiency of its efforts by applying a broader analytical approach—called "integrated safeguards"— that could result in reduced inspections in those countries that had a proven record of compliance with their safeguards obligations. The concept of the integrated safeguards approach was for inspectors to expand on the traditional material accountancy approach by being "less checklist-oriented" and more analytical and, therefore, comprehensive in their country evaluations. IAEA's implementation of integrated safeguards, however, is applicable only to countries that have both a comprehensive safeguards agreement and an Additional Protocol in force and for which IAEA has been able to draw a "broader conclusion." A broader conclusion means that, for a given year, a country has demonstrated that all declared nuclear material remained in peaceful activities and there are no indications of undeclared nuclear activities. In our 2005 report, we also found that another safeguards protocol that applies to countries with small quantities of nuclear material—called the small quantities protocol—limited IAEA's ability to detect undeclared nuclear material or activities.[19] Specifically, this protocol holds in abeyance certain safeguards measures as long as a country meets specific eligibility requirements. Countries may request that the small quantities protocol, which is subject to Board of Governors approval, supplement their comprehensive safeguards agreement.

Table 1. IAEA Safeguards and Nuclear Security Programs Budget for 2011

Dollars in millions			
Program	**Regular budget, as reported by IAEA**	**Extra-budgetary cash contributions, as reported by IAEA**	**U.S. extra-budgetary cash contributions, as reported by State**
Safeguards program	$166.1	$39.3	$24.8[a]
Nuclear security program	$5.3	$25.0	$13.4[b]

Source: GAO presentation of IAEA and State budget data.

Note: Regular budget and extra-budgetary cash contributions were calculated at the average exchange rate used by the United Nations in 2011 of $1.3893 to €1.00.

[a] This figure includes extra-budgetary cash contributions from State ($24.2 million) and NNSA ($0.6 million).

[b] This figure includes extra-budgetary cash contributions from State ($7.0 million), NNSA ($6.1 million), and the Department of Homeland Security ($0.3 million).

Nuclear Security Program

According to IAEA's *Nuclear Security Plan 2010-2013*, the risk that nuclear or other radioactive material could be used in malicious acts remains high and is regarded as a serious threat to international peace and security. For example, since the creation of IAEA's Incident and Trafficking Database in 1995 (formerly called the Illicit Trafficking Database), member countries have reported over 2,300 incidents involving nuclear and other radioactive materials detected outside of regulatory control, including incidents of illicit trafficking. IAEA also projects a net increase in the number of nuclear reactors in operation by 2030. As a result, IAEA has augmented its efforts for countries that request its assistance in protecting their nuclear and radiological materials and facilities.[20] As we previously reported, in the hands of terrorists, these materials could be used to produce a simple and crude but potentially dangerous weapon, known as a dirty bomb, by packaging explosives with the radioactive material for dispersal. Through its Office of Nuclear Security, IAEA provides advisory services to member countries to establish the necessary infrastructure to protect nuclear and other radioactive materials from theft and diversion, protect nuclear installations and transport against sabotage and other malicious acts, and combat illicit trafficking in nuclear and other radioactive materials. IAEA also disseminates information about procedures for dealing effectively with nuclear threats through international conferences, training courses, and security guidance publications. In addition, IAEA assists member countries in implementing nuclear security measures at major public events, such as the 2011 Pan American Games in Mexico and the 2012 European Football Championship in Poland and Ukraine.

Fuel Bank Program

IAEA's international nuclear fuel bank is a relatively new program, in contrast to the safeguards and nuclear security programs. The purpose of the fuel bank is not to provide a routine supply of LEU fuel but to guarantee an international supply for civilian nuclear power in the event of a supply disruption. As specified by the agency's Board of Governors, the fuel may be purchased on a non-discriminatory, non-political basis by IAEA member countries that are meeting their nonproliferation obligations with the agency. These obligations include having a comprehensive safeguards agreement in force. The fuel bank is also meant to provide countries with a "credible alternative" to the pursuit of sensitive nuclear fuel enrichment activities that employ the same technology that can be used to produce nuclear material for nuclear weapons. One of the Board of Governors' conditions for the fuel bank is that it not disrupt the international nuclear fuel market. Funding for the fuel bank has come from private and public sources. In addition to the U.S. government's extra-budgetary cash contribution of nearly $50 million, a Washington, D.C.-based nongovernmental organization—the Nuclear Threat Initiative—has provided $50 million,[21] and other donors have collectively contributed or pledged nearly $60 million.[22] In December 2010, IAEA's Board of Governors approved the establishment of the fuel bank with the support of 28 of the 35 board members, including the United States.

IAEA HAS STRENGTHENED ITS SAFEGUARDS PROGRAM IN RECENT YEARS, BUT CHALLENGES MAY LIMIT EFFORTS TO FURTHER IMPROVE EFFECTIVENESS

IAEA has taken four key steps to strengthen its safeguards program since we last reported on this in 2005: (1) successfully encouraged more countries to bring an Additional Protocol into force, giving IAEA expanded authority and tools to detect undeclared nuclear activities; (2) increased the number of countries subject to a broader range of safeguards measures; (3) started to upgrade its safeguards analytical laboratories in Austria with new equipment and facilities; and (4) produced its first long-term strategic plan for the Department of Safeguards to formally identify program objectives and goals. IAEA plans to further improve the effectiveness and efficiency of the safeguards program by expanding implementation of the "state-level concept" to all countries with safeguards agreements by 2014. The state-level concept is an approach in which the agency considers a broad range of information about a country's nuclear capabilities and tailors its safeguards activities in each country accordingly. However, several member countries have raised concerns about the state-level concept that may hinder broader implementation. In addition, the agency has not quantified the resources that it may need to expand implementation of the state-level concept, making it difficult to determine the associated long-term costs and benefits.

More Countries Have Concluded an Additional Protocol, Giving IAEA Expanded Authority

Since 2005, 46 more countries have brought an Additional Protocol into force, which gives IAEA greater authority and tools to search for undeclared nuclear facilities and activities. Specifically, 119 countries had concluded an Additional Protocol as of the end of December 2012, compared with 73 countries at the end of December 2005.[23] Moreover, since our 2005 report, the United States and Russia were the last of the five nuclear weapon states to bring an Additional Protocol into force, which agency officials told us has helped make IAEA's outreach efforts to encourage other countries to conclude an Additional Protocol more effective and credible.[24] As a result of more countries bringing an Additional Protocol into force, IAEA has significantly increased the number of countries for which the agency has been able to draw broader conclusions about both declared and undeclared activities over the past 7 years. In 2011, IAEA drew broader conclusions for 58 non-nuclear weapon states, and evaluations regarding the absence of undeclared nuclear material were still ongoing for the other 51 countries that had both comprehensive safeguards agreements and Additional Protocols in force at that time.[25] By comparison, at the end of 2005, IAEA had drawn broader conclusions for 24 non-nuclear weapon states and had such evaluations in progress for 46 other countries. A broader conclusion refers to the agency's determination that, for a given year, a country has demonstrated that all declared nuclear material remained in peaceful activities and there are no indications of undeclared nuclear activities. IAEA can draw a broader conclusion only for those countries that have both a comprehensive safeguards agreement and an Additional Protocol in force. By contrast, without an Additional Protocol in force, the agency cannot provide credible assurances regarding the absence of undeclared nuclear material and activities in the country as a whole. In such cases, IAEA reports its conclusion for a given year only with respect to whether declared nuclear material remained in peaceful activities. Figure 1 shows the number of countries with or without an Additional Protocol in force, by year, from 2005 through 2012. Appendix II lists the status of countries with safeguards agreements and an Additional Protocol in force as of the end of December 2012.

To encourage more countries to bring an Additional Protocol into force, the agency continued to implement a plan of action, developed in 2001 and updated every year, by intensifying its efforts to conclude Additional Protocols, especially with countries that have significant nuclear activities. In particular, IAEA has continued its efforts to hold training events and workshops at the national, regional, and international levels with countries to explain what their obligations would be under an Additional Protocol. For example, the agency held four outreach events in 2011, including a regional seminar for southeast Asian countries with significant nuclear activities and a number of bilateral briefings in New York and Geneva.

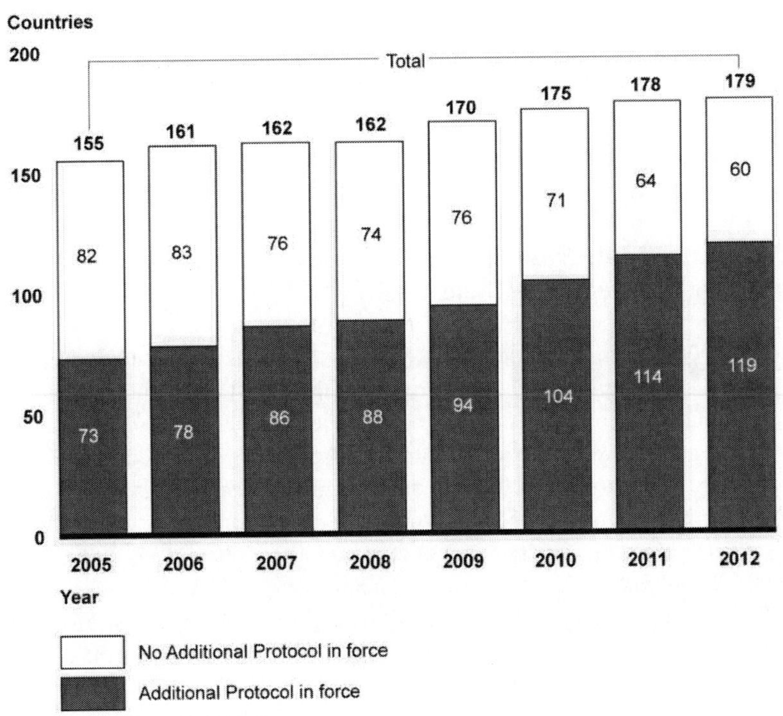

Source: GAO presentation of IAEA data.

Note: The total number of countries with or without an Additional Protocol in force from 2005 through 2012 reflects the number of countries that had safeguards agreements in force with IAEA during those years. Data do not include North Korea, as IAEA has not been able to conduct all necessary safeguards activities in that country since 1994. Data also do not include Taiwan, although Taiwan has had a safeguards agreement in force since 1969 and has implemented the measures of the Model Additional Protocol since October 1999.

Figure 1. Countries with or without an Additional Protocol in Force, from 2005 through 2012.

While universal adoption of the Additional Protocol remains a goal for the agency, many countries have not brought it into force. According to a public statement by the Head of IAEA's Department of Safeguards, of the 60 countries with safeguards agreements that have not concluded an Additional Protocol as of December 2012, 7 are engaged in significant nuclear activities: Algeria, Argentina, Brazil, Egypt, Iran, Syria, and Venezuela. Of those seven countries, Algeria received approval from IAEA's Board of Governors for an Additional Protocol in September 2004 but has not signed it or brought it into force, Iran signed an Additional Protocol in December 2003 but has not brought it into force, and the other countries have taken no official action concerning an Additional Protocol.[26] As we reported in 2005, some countries may be hesitant to conclude an Additional Protocol due to (1) concerns about the financial costs associated with it; (2) an unwillingness to submit to an intrusive inspection regime; (3) their inadequate regulatory systems to collect information on all nuclear-related activities; or (4) political factors, such as the policies of neighboring countries or the time needed for national legislatures to approve an Additional Protocol. For instance, Brazil has continued to take the position that an Additional Protocol does not need to be universally applied to all countries and that inspections and IAEA requests for information are overly intrusive. Brazil has also stated that it will not undertake further nonproliferation obligations, including adoption of the Additional Protocol, until progress has been made by nuclear weapon states on disarmament. In addition, Egypt will not bring an Additional Protocol into force until Israel signs the NPT, concludes a comprehensive safeguards agreement, and implements an Additional Protocol.

IAEA Has Increased the Number of Countries That Are Subject to a Broader Range of Safeguards Measures

Since our 2005 report, IAEA has (1) tightened the terms and conditions of a protocol brought by many countries into force concerning small quantities of nuclear material and (2) increased the number of countries that have comprehensive safeguards agreements in force. As a result, the agency has increased the number of countries to which it applies a broader range of safeguards measures. Specifically, in September 2005, IAEA's Board of Governors approved revisions to the text of the small quantities protocol that, when in force, holds in abeyance certain safeguards measures as long as a country meets specific eligibility requirements. As revised in 2005, the small quantities protocol is no longer available to a country with either an existing or planned nuclear facility, and it has reduced the number of safeguards measures held in abeyance. For example, the revised small quantities protocol obligates countries to make initial declarations to IAEA on the nuclear materials they hold, no matter how small, and reinstates the agency's right to conduct ad hoc and special inspections. Since 2005, IAEA has worked with countries to voluntarily amend or rescind their original small quantities protocols to increase the agency's ability to detect undeclared nuclear material or activities. As of the end of December 2012, 46 countries had brought the revised small quantities protocol into force, including 34 countries that amended their original small quantities protocols to reflect the revised text; 4 countries had rescinded their original small quantities protocols; and 52 countries still had the original small quantities protocol in force. (App. II lists the status of countries with small quantities protocols in force as of December

2012.) By contrast, as of July 2005, 76 countries had the original small quantities protocols in force.

In addition, IAEA has continued to promote the importance of comprehensive safeguards agreements through seminars, briefings, and other outreach efforts—resulting in 24 more countries bringing such agreements into force since our 2005 report. Specifically, as of December 2012, 171 of the 184 non-nuclear weapon state parties to the NPT had brought comprehensive safeguards agreements into force,[27] compared with 147 countries at the end of 2005. The NPT requires non-nuclear weapon states to conclude comprehensive safeguards agreements within 18 months of becoming a party to it.[28] However, as we previously reported, IAEA officials have indicated that several countries have not yet brought such agreements into force because they consider it to be a low priority compared with other national priorities or because of other factors.

IAEA Has Started to Upgrade Its Safeguards Analytical Laboratories

To strengthen its capabilities to provide independent and timely analysis of environmental and nuclear material samples, IAEA embarked on an estimated $112 million project in 2007 to upgrade its safeguards analytical laboratories in Seibersdorf, Austria, with new equipment and facilities. These improvements are expected to result in higher-quality and faster analyses of samples collected by IAEA inspectors, particularly as the agency anticipates that its workload will rise with a projected expansion of civilian nuclear power in the next 30 years. As one of its measures to verify that countries' nuclear material declarations are correct and complete, the agency relies on its facilities in Seibersdorf and other laboratories in a global network to analyze hundreds of samples annually. The modernization of IAEA's Seibersdorf laboratories is the largest capital improvement project the agency has ever undertaken. IAEA's public statements indicate that the project has involved forecasting and defining analytical and infrastructure needs, designing the laboratories, and site infrastructure and security improvements. In the first phase of this two-phase project, the agency expanded its Environmental Sample Laboratory in 2011 and installed additional equipment to enhance its ability to detect undeclared nuclear material or activities. According to IAEA, this equipment is capable of finding "a needle in a haystack" by measuring minute particles of uranium found in environmental samples. Figure 2 shows the upgraded equipment in service at IAEA's newly expanded Environmental Sample Laboratory. In addition, when we visited the laboratory in June 2012, IAEA officials told us that the newly expanded facilities will significantly help prevent potential contamination of samples, which was a concern at the older facilities due to cramped workspace and inadequate air filtration. In the project's second phase, construction for a new facility to house IAEA's Nuclear Material Laboratory in Seibersdorf began in 2012 and is scheduled for completion in 2014.

Although the project is being partially funded by the agency's regular budget ($32 million through 2013), IAEA also solicited and received significant extra-budgetary contributions from several member countries and the European Commission. As of early December 2012, the United States was the biggest donor, having contributed $21.0 million, and other donors had collectively pledged or contributed $30.1 million.[29] At that time, the agency planned to receive another $11.9 million in funding and still needed to raise $17.3

million to complete the project. Some experts told us the agency had made a compelling case for the project by demonstrating the need to member countries to replace facilities and equipment that were outdated and deteriorating, putting the agency's ability to provide independent and timely analysis at risk.

In addition to modernizing its Seibersdorf facilities, IAEA has continued its efforts to increase the number of laboratories it has qualified to be part of its global Network of Analytical Laboratories. The expansion of the network is intended to improve the agency's capacity to handle environmental and nuclear material samples and reduce processing times. The agency typically relies on two network laboratories to analyze a given sample to ensure accuracy, based on consistent results. In 2005, we reported that the network was being used beyond its capacity, resulting in untimely processing and reporting of results—averaging 2 to 3 months and sometimes taking as long as 6 months. At that time, in addition to the agency's own laboratories, there were 15 laboratories in 7 member countries (including the United States) and the European Commission in the network. As of July 2012, the network had expanded to comprise 19 laboratories in 8 member countries and the European Commission, in addition to the agency's own laboratories. IAEA also reported, as of July 2012, that it was in the process of reviewing qualifications for additional laboratories in Argentina, Belgium, China, Finland, France, Germany, Hungary, South Korea, and the United States to join the network.

Source: © May 2011 Klaus Gaggl, IAEA.

Figure 2. Upgraded Equipment at IAEA's Environmental Sample Laboratory in Seibersdorf, Austria.

Safeguards Department Produced Its First Long-Term Strategic Plan

In August 2010, after conducting an 18-month internal review, IAEA's Department of Safeguards produced a long-term strategic plan for the safeguards program for 2012 through 2023. To date, this represents the first long-term strategic planning effort by any IAEA department. While the agency as a whole does not have a long-term strategic plan in place, officials from State and Brookhaven National Laboratory credited the Department of Safeguards for developing what they described as a forward-thinking plan.[30] According to a report by one expert we interviewed, "The exercise is unprecedented for the Agency and virtually unheard of in [United Nations]-related organizations."

IAEA has not publicly released the full long-term safeguards plan, therefore we did not evaluate it. However, agency officials have made presentations and released a 10-page summary of the plan. The summary identifies three objectives, seven strategies and goals, and dozens of supporting activities covering (1) the conceptual framework for safeguards implementation, (2) legal authority, (3) technical capabilities, (4) cooperation and partnerships, (5) human resources, (6) financial resources, and (7) communication. For example, to support the financial resources strategy and the goal of maintaining member countries' confidence in the cost-effective implementation of the safeguards program, the plan states that the department will

- assess budgetary needs and seek sufficient and predictable funds and prepare for unexpected expenses;
- optimize the use of financial resources and seek further efficiencies through, for example, cost-benefit analyses to guide decision-making; and
- monitor and measure performance toward the achievement of desired results and increased transparency and accountability in reporting to member countries.

According to IAEA presentations, the expected benefits of the long-term safeguards plan include promoting strategic thinking and preparing the department for future challenges; unifying departmental planning, monitoring, and evaluation activities under one framework; supporting internal decision-making concerning priorities and resource allocation; and providing better information to member countries. Agency documents also indicate that the department will adjust the long-term safeguards plan every 2 years, if necessary, following a review of the external environment and a risk reassessment. In addition, the department is to update the long-term safeguards plan every 6 years, based on a performance assessment, and will provide input to the development of the next agency-wide, medium-term strategy.

IAEA Plans to Expand Implementation of the State-Level Concept by 2014 to Better Allocate Safeguards Resources

IAEA plans to further strengthen the effectiveness and efficiency of its safeguards program by expanding the implementation of an approach called the state-level concept to all countries with safeguards agreements. According to IAEA documents and officials, this next phase in the evolution of safeguards would allow the agency to apply a tailored set of safeguards activities and allocate resources for each country as appropriate, based on a range of factors about countries' nuclear activities and capabilities as a whole. Such factors include a country's nuclear fuel cycle facilities, knowledge and expertise, legal framework, domestic safeguards authority, and past research and development. The agency first introduced the state-level concept about 10 years ago under integrated safeguards—an approach that encouraged inspectors to be more analytical and comprehensive in their country evaluations. Since that time, IAEA has implemented the state-level concept only in those countries that have both a comprehensive safeguards agreement and an Additional Protocol in force and for which the agency has drawn a broader conclusion that all nuclear material remains in peaceful activities. As of mid-2012, the agency reported it had implemented state-level approaches for 53 such countries, plus Taiwan. However, IAEA documents indicate that broader implementation of the state-level concept would yield greater efficiencies in allocating more resources to any issues of safeguards concern. Consequently, the agency plans to apply the state-level concept to all countries under safeguards, regardless of whether they have a comprehensive safeguards agreement and Additional Protocol in force, by 2014.

Agency documents also indicate that broader implementation of the state-level concept will affect IAEA's planning, conduct, and evaluation of safeguards activities for each country. Specifically, IAEA indicates it will consider a complete range of factors about countries' nuclear activities to determine the way in which the agency implements safeguards activities for each country and allocates resources in the field, where inspections are conducted, and at IAEA headquarters, where open-source, remote monitoring, and other analysis is conducted. For example, IAEA officials told us that the agency's safeguards approach might differ for a country with few nuclear facilities and a strong domestic safeguards authority, compared with the approach used for a country with many facilities and a weaker safeguards authority. Ultimately, IAEA officials told us that broader implementation of the state-level concept would allow the agency to better allocate resources by reducing safeguards activities in those countries for which the agency has been able to establish a complete picture of nuclear activities, including a determination that there is no indication of undeclared activities, thus allowing the agency to focus greater efforts and resources on any issues of safeguards concern. Furthermore, in contrast to the annual evaluation process used in the past—including under the integrated safeguards approach—the agency's evaluations of a country under broader implementation of the state-level concept would be continuous and ongoing, and the safeguards approaches in a given country would be modified accordingly. According to agency documents, this approach would ensure the agency's conclusions remain soundly based and up-to-date. Figure 3 shows IAEA's process for evaluating countries' safeguards compliance using the state-level concept.

According to public statements by the Head of IAEA's Department of Safeguards over the past 2 years, it is necessary to broaden implementation of the state-level concept because past safeguards approaches have been insufficient to detect undeclared activities, such as a clandestine nuclear weapons program in Iraq and covert supply networks in Libya.[31] In addition, the absence of an Additional Protocol in force in countries such as Iran and Syria has restricted the agency's ability to look beyond declared facilities and activities, preventing the agency from drawing a conclusion on whether there is undeclared nuclear material or activities in those countries. The Head of Safeguards has also publicly stated that past safeguards approaches have been too "narrow, prescriptive, and criteria-driven"—based on the nature and quantity of nuclear material and focused on specific safeguards activities for different types of facilities within a given country—rather than looking at the country more comprehensively. As a result, the Head of Safeguards has said, "satisfying the criteria" is in danger of becoming the objective, rather than addressing the underlying purpose of the work—to determine whether there is any indication of nuclear proliferation. According to this senior IAEA official, previous cases of safeguards violations or clandestine nuclear weapons development occurred in countries believed to have limited nuclear fuel cycles and involved previously exempted or undeclared nuclear material.[32] Given these previous cases, the Head of Safeguards has stated, the quantity of nuclear material in a country's possession, or even the size of its fuel cycle, is not necessarily a sufficient indicator of proliferation risk. Instead, IAEA needs to implement safeguards in a more adaptable, less predictable manner that considers the country's activities as a whole.

In addition to improving the effectiveness of its safeguards program, IAEA officials told us that broader implementation of the state-level concept would improve the program's efficiency by reducing routine inspections where IAEA has found no indication of undeclared activities and by directing resources to any issues of safeguards concern. State-level concept proponents told us that even those countries where the integrated safeguards approach has been applied in recent years could benefit from a further reduced inspection burden, assuming those countries remain compliant with their safeguards obligations. In particular, Japan, Germany, and Canada have traditionally comprised the bulk of IAEA safeguards inspections, in terms of time and cost, based on the number and type of declared nuclear facilities in those countries. The number of inspections has declined in those three countries under integrated safeguards since our 2005 report; for example, inspections in Canada declined from a peak of 1,337 person-days of inspection[33] in 2004 to 436 person-days of inspection in 2011. Japan, Germany, and Canada have expressed their support for the state-level concept.

Several Countries Have Raised Concerns That May Hinder Broader Implementation of the State-Level Concept

Several member countries, including the United States, support IAEA's plans to broaden implementation of the state-level concept, but other member countries—including some countries with significant nuclear activities—have raised concerns that the agency has not clearly defined and communicated how the state-level concept will be implemented or how it will stay within bounds of the agency's existing legal authorities. Consequently, if these concerns are not resolved, IAEA may not be able to implement the state-level concept in all countries with safeguards agreements by 2014. For example:

- Officials from some foreign missions in Vienna, Austria, told us that they believe the state-level concept does not have objective criteria and therefore IAEA could implement it in an inconsistent, subjective, and potentially discriminatory manner. For instance, officials from one foreign mission told us that IAEA needs to better explain its vision for the state-level concept in a clear, transparent, and specific manner to obtain broader acceptance among member countries. Officials from another foreign mission told us that they are concerned about IAEA's ability to draw safeguards conclusions without clearly defined criteria in place under the state-level concept.

- Officials from several foreign missions also told us that a number of countries are concerned that the state-level concept may give the agency too much latitude in gathering information or allow the agency to be too intrusive into a country's civilian nuclear operations. For example, officials from one foreign mission stated that it is important for IAEA to rely on verifiable information that is relevant only to safeguards, whereas IAEA has stated that it would use "all available sources of information" about a given country. Officials from another foreign mission emphasized that the state-level concept cannot be "a blank check for the agency to do whatever it wants" in a country and that it is important for the agency to avoid unnecessarily disrupting nuclear power plant operations in a country "just because the agency has ill-defined criteria."
- Officials from one foreign mission also observed that a lack of clarity about the state-level concept could undermine the agency's efforts to encourage more countries to conclude an Additional Protocol because those countries may be suspicious of how the agency would use the additional access and information granted by an Additional Protocol.

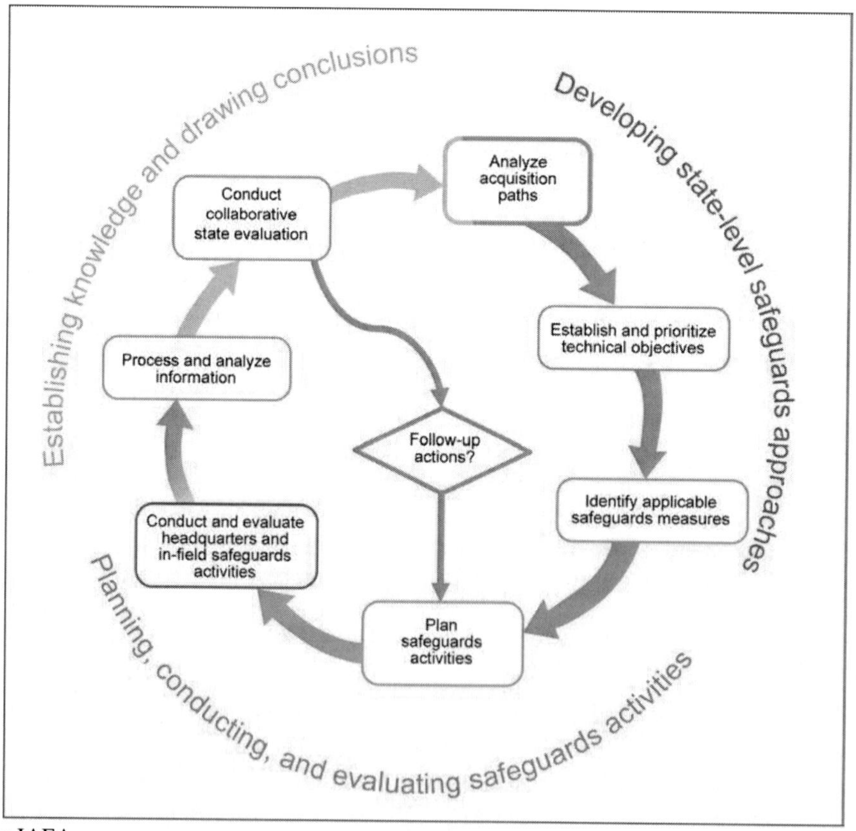

Source: IAEA.

Figure 3. IAEA's Process for Evaluating Countries' Safeguards Compliance Using the State-Level Concept.

As of April 2013, IAEA had not developed the criteria or communicated specific details to countries about how the state-level concept would be implemented. Furthermore, IAEA officials told us they did not intend to communicate specific implementation details to countries. Officials from State, which coordinates the United States' financial and policy relationship with IAEA, told us that IAEA has been transparent in developing the state-level concept and has repeatedly informed the Board of Governors about its progress. In addition, State maintains that, as a matter of principle, IAEA's implementation of safeguards should remain impartial, objective, based on relevant information, and rooted in the agency's existing legal framework. Nonetheless, due to the concerns raised by several countries, the General Conference passed a resolution in September 2012 that included a request for IAEA's Secretariat to report to the Board of Governors on the conceptualization and development of the state-level concept. A senior IAEA official told us that the agency expects to report to the Board of Governors in 2013. In the meantime, according to agency officials, IAEA has continued its efforts to communicate the state-level concept with countries and other stakeholders through multiple forums, including conferences, presentations, and bilateral discussions. During our review, we also observed that IAEA modified its description of the state-level concept to address concerns. For example, in the agency's annual report for 2010, a paragraph describing safeguards conclusions states that "the safeguards system being implemented is described as 'information-driven.'" Since that time, IAEA has moved away from the term "information-driven safeguards" to avoid language that could suggest inappropriate information-gathering. In addition, IAEA officials have emphasized that they are putting in place guidance to ensure consistency under the state-level concept in how the agency (1) evaluates individual countries' nuclear activities and (2) implements appropriate safeguards in each country. To ensure that IAEA's evaluations of individual countries' nuclear activities under the state-level concept are sound and independent of individual analysts' judgment, officials from the Department of Energy's Lawrence Livermore National Laboratory told us that IAEA's internal review process appears to be placing greater emphasis on information-sharing, collaboration, and rigorous review than in the past.[34] Nonetheless, IAEA officials stated that no matter how they articulate what the state-level concept is, certain countries may never support it.

IAEA Has Not Quantified the Long-Term Costs and Benefits of Broadly Implementing the State-Level Concept

IAEA reports annually on its progress in utilizing various resources to strengthen the safeguards program, but the agency has not quantified the additional human capital, technological, or other resources that may be needed to broaden implementation of the state-level concept. As a result, it is difficult to determine the long-term costs and benefits associated with the state-level concept, and the agency may not be able to build the necessary political and, potentially, financial support to implement the state-level concept in all countries with safeguards agreements by 2014. For example, in recent annual reports and other documents, IAEA has reported progress in training its inspectors and analysts and integrating its information systems to enhance the agency's ability to evaluate a full range of information about countries' nuclear activities and to draw soundly based conclusions about countries' compliance with their safeguards obligations. However, the agency has not quantified the extent to which this integration has already been achieved or detailed how and when it will address any remaining expertise, technology, or other resource gaps. Similarly, IAEA reports annually on the number of cameras, unattended radiation monitors, and other remote monitoring systems it operates for safeguards activities; however, the agency does not indicate the cost or specific benefits associated with these items. A senior IAEA official told us that it is difficult, if not impossible, to quantify the long-term benefits of implementing the state-level concept, but that it should result in (1) reduced safeguards activities in countries with little or no issues of safeguards significance and (2) increased efforts in countries with any such issues. This official added that IAEA's safeguards activities for an individual country may change as issues and circumstances change from year to year.

Nonetheless, we and others have previously reported on the need for IAEA to identify costs and manage budget requirements. In IAEA's financial statements for 2011, the agency's external auditor also found that IAEA "has not been able to identify savings in a transparent and comprehensible manner" in the 10 years since the agency took additional steps to strengthen its safeguards program with the integrated safeguards approach. The external auditor stated that to measure improvements, the resources needed to launch and operate a project—as well as the benefits for member countries and the agency—should be measurable. In 2005, we also reported that IAEA does not have a process in place to systematically evaluate its safeguards resource requirements over the long-term because the agency's budget process and research and development plan (a document now called the development and implementation support program for nuclear verification) are designed to forecast resource needs on a 2-year basis.[35] At that time, we recommended that State work with the agency to develop a systematic process to forecast long-term safeguards budget requirements beyond the 2-year budget cycle. State agreed with our recommendation, and IAEA has taken some steps to address it. During this review, we observed that the Department of Safeguards' current development and implementation support program document for 2012-2013 identifies 24 specific projects that the department wants to undertake and fund in the current, 2-year budget cycle, mostly through member countries' extra-budgetary contributions. These projects are intended to meet specific objectives identified in the department's *Long-Term Strategic Plan 2012-2023* and, according to a senior IAEA official, they involve tasks that are not addressed through the agency's routine safeguards activities. However, the document does not identify monetary values for these various projects or take a longer-term approach to quantify the resources needed to broaden implementation of the state-level concept.

Given the current zero-real-growth budget policy environment, IAEA officials stated that the state-level concept may not yield overall financial savings but would allow the more effective use of resources by shifting existing resources toward any issues of safeguards concern. Nonetheless, without a clear and concise assessment of the resources and the associated long-term costs and benefits, including trade-offs— such as reduced inspections for increased analytical capabilities— member countries and the international community at large may have difficulty understanding the benefits of the state-level concept. Furthermore, if the agency conducted an assessment of its long-term resource needs and determined that full implementation of the state-level concept warrants a greater investment than the current budget provides, the agency could present a credible foundation from which to request additional resources from member countries.

IAEA Has Continued to Support Nuclear Security, but Three Key Issues Limit the Agency's Ability to Ensure That Its Resources Are Used Efficiently and Effectively

Since our last report in 2005, IAEA has continued to help countries improve the security of their nuclear material and facilities, but three key issues limit the agency's ability to ensure that its nuclear security resources are used efficiently and effectively. First, IAEA's heavy reliance on extra-budgetary contributions for its nuclear security activities makes it difficult to plan and implement projects because these funds vary from year to year and because donor countries often place conditions on their contributions. Second, IAEA has not conducted a needs-based assessment of the resources required beyond its 2-year budget cycle, which also hinders its ability to ensure that resources are directed to the greatest security needs. Third, IAEA does not systematically report the results of performance measures related to its nuclear security program, thus the extent to which the agency is meeting its nuclear security goals is unclear.

IAEA Continues to Support Countries' Nuclear Security Efforts through Its Nuclear Security Program

Since our 2005 report, IAEA's role in nuclear security has gained increased prominence and has been reinforced through international Nuclear Security Summits held in 2010 and 2012. Specifically, participants at the 2010 Nuclear Security Summit in Washington, D.C., issued a statement recommending, among other things, that countries (1) work with IAEA to implement the agency's guidance on nuclear security and (2) request the agency's assistance in assessing the physical protection systems of civilian nuclear material and facilities. Two years later, participants at the 2012 Nuclear Security Summit in Seoul, South Korea, issued a statement that highlighted the importance of IAEA's role in nuclear security, encouraging countries to contribute funds toward the agency's efforts and supporting the agency's sponsorship of a conference in 2013 to coordinate nuclear security activities.

In addition, IAEA has enhanced its training activities and continued other activities since our 2005 report—such as providing advisory services and managing a database on incidents involving nuclear and other radioactive materials detected outside of regulatory control—to support countries' efforts to improve the security of their nuclear material and facilities. While security of nuclear and radiological materials is first and foremost a national responsibility, IAEA responds to countries' requests to participate in the agency's voluntary nuclear security program. According to NNSA officials, one of IAEA's most important functions in the area of nuclear security is the range of training activities it offers to countries on physical protection of nuclear facilities and materials. In particular, NNSA officials told us that training is an essential component to countries developing and sustaining a national nuclear security regime, as it helps build a cadre of capable personnel who are responsible for maintaining and enhancing a country's nuclear security infrastructure. The Director of IAEA's Office of Nuclear Security also told us that it is equally important for countries to have a minimum number of well-educated, trained people as it is to have radiation detection equipment. Therefore, IAEA offers a wide variety of international, regional, and national training courses and workshops. We reported in 2005 that IAEA had conducted almost 80 training courses for about 1,500 participants from January 2001 through March 2005.[36] In IAEA's *Nuclear Security Report 2012*, the agency stated that it conducted 70 nuclear security training programs and workshops that reached over 1,750 people from July 2011 to June 2012—indicating that the agency is now reaching more people in a shorter period of time. In addition, since 2005, IAEA has sponsored several new training initiatives in support of nuclear security efforts, including

- completing a nuclear security training site in Obninsk, Russia—the agency's largest project to date on physical protection—which offers in-field training and hands-on exercises in nuclear security;
- developing the International Network for Nuclear Security Training and Support Centers, designed to facilitate human resource development and provide technical support services, and;
- creating the International Nuclear Security Education Network, which provides a forum for IAEA, educational institutions, and research bodies to collaborate on nuclear security education. The network is open to any educational and research institution already involved or that plans to be involved in nuclear security education in the future.

IAEA has also continued to provide International Physical Protection Advisory Service (IPPAS) missions to countries that request them. These voluntary missions assist countries in strengthening their national civilian nuclear security regimes by providing (1) guidance on the protection of nuclear material and facilities, as well as sealed radioactive sources and other radioactive material, (2) best practices in nuclear security, and (3) peer advice on implementing international agreements related to physical protection of nuclear material and facilities. The 2010 Nuclear Security Summit recognized the value of these missions, while the 2012 summit encouraged countries to enhance their physical protection and welcomed IAEA's assistance in these efforts. The Director of the Office of Nuclear Security told us that, as of June 2012, the agency had conducted 57 IPPAS missions in 37 countries and that the number of countries requesting these missions continues to grow. During an IPPAS mission, at the request of a country, IAEA assembles a team of international experts that assesses the country's system of physical protection, compares it with international guidelines and best practices, and makes recommendations for improvements to correct any identified deficiencies. IAEA's nuclear security guidelines establish the standard by which the United States and other countries generally classify the categories of physical protection that should be given to nuclear material, based on the type, volume, and disposition of the material. The results of IPPAS missions, which are contained in a report provided to each country, are confidential. According to NNSA officials, the agency believes it is obligated to keep the information confidential to respect countries' concerns that security vulnerabilities in the reports could be made public. To promote the continued use of IPPAS missions, IAEA does not share the reports with the public or other countries. In addition, countries that agree to host an IPPAS mission are not required to implement any recommendations contained in the mission report. However, IAEA conducts follow-up visits to determine whether the country has implemented the agency's recommendations.

IAEA has also increased its efforts to assist countries in improving their capabilities to detect, track, and respond to illicit trafficking of nuclear and radiological materials. In 2005, we reported that IAEA was encouraging countries to contribute data to its Incident and Trafficking Database (formerly called the Illicit Trafficking Database) to help identify trends in incidents involving nuclear and radiological materials detected outside regulatory control.[37] Since that time, the agency has increased the number of countries participating in the database by nearly 50 percent: from 81 countries in June 2005 to 121 countries as of February 2013. Using this data, the agency produces quarterly and annual reports to facilitate information exchange on trends related to incidents involving unauthorized acquisition, provision, possession, use, transfer, or disposal of nuclear and other radioactive materials. In addition, since 2007, IAEA has held 14 regional information meetings worldwide to, among other things, (1) help raise awareness of the database, (2) improve the quality and accuracy of reporting in the database, and (3) highlight the support that IAEA can offer to countries to improve their nuclear security efforts. Participants from over 100 countries have attended one or more of these meetings. In July 2012, IAEA also held a meeting with representatives from 81 countries to discuss, among other things, how to use information from the database in ways that would further satisfy their security needs.

Three Key Issues Limit IAEA's Nuclear Security Program

IAEA's nuclear security program faces three key issues that limit the agency's ability to ensure that its resources are used efficiently and effectively:

- *IAEA relies heavily on extra-budgetary contributions for funding its nuclear security activities.*[38] Certain member countries, particularly the United States, and other donors provide IAEA with extra-budgetary contributions on a voluntary basis to supplement the agency's regular budget and to meet critical needs. These extra-budgetary contributions vary from year to year and, according to the Director of the Office of Nuclear Security, are often subject to donors' conditions. As a result, it is difficult for the agency to plan and implement projects where security needs may be greatest. This heavy reliance on extra-budgetary contributions is a long-standing challenge that we reported on in 2005. At that time, we reported that 89 percent of the agency's total nuclear security funding came from extra-budgetary contributions, and the remaining 11 percent came from the agency's regular budget and was used primarily for staff salaries. We recommended that IAEA consider determining whether the nuclear security program receives adequate regular budget funds.[39] In February 2010, a State official reported that IAEA had made efforts in this regard by moving toward a model in which administrative overhead costs, such as salaries, are paid from the regular budget, and special projects are funded through extra-budgetary contributions. However, during this review, we found that IAEA continues to rely heavily on extra-budgetary contributions, accounting for $25 million or 82.5 percent of its total $30.3 million funding for nuclear security in 2011.[40] The majority of this extra-budgetary funding is designated by the donor countries for specific projects. According to the Director of the Office of Nuclear Security, the conditions that donor countries often place on their extra-budgetary contributions limit the agency's ability to establish its own priorities and ensure that the resources are directed to the areas of greatest need. For example, officials from one foreign mission told us that their country had contributed $8 million over the past 4 years and designated it for specific projects, such as improving the security of radiological materials at hospitals or providing nuclear security training, in certain countries.

- *IAEA has not conducted a needs-based assessment of the resources required for the nuclear security program beyond the agency's 2-year budget cycle.* Consequently, IAEA has not identified how it will meet any long-term resource needs through the regular budget, extra-budgetary contributions, or both. The absence of such information hinders the agency's ability to ensure that resources are directed to the greatest security needs. We have previously reported that good strategic planning practices include describing the relationship between long-term goals and budgetary needs.[41] IAEA's Director of the Office of Nuclear Security told us that the agency has been working for the past year to identify countries' needs through individual nuclear security support plans. The objectives of these plans are to (1) identify and consolidate a country's nuclear security needs into a single document that includes the necessary security improvements, as well as an appropriate framework for implementing these improvements, and (2) to quantify the resources needed and prioritize the agency's activities to meet those needs. According to the director, these individual plans are designed to identify actions required and approved by both the country and IAEA to ensure that the country's nuclear security regime is effective and sustainable. The director told us that the agency expects to have plans in place for over 100 countries by July 2013 and intends to use this approach for short-, medium-, and long-term assessment of countries' needs and the agency's program and budget priorities. The director added that these plans would also inform the agency's upcoming *Nuclear Security Plan 2014-2017*. Nonetheless, the agency does not have an overall long-term plan that incorporates the country-specific plans and looks comprehensively at nuclear security needs across all countries and identifies resources and priorities for the agency.
- *IAEA does not systematically report the results of performance measures related to its nuclear security program.* As a result, the extent to which the agency is achieving its nuclear security goals is unclear. IAEA has taken steps in recent years to develop measures for assessing the performance of its nuclear security efforts, but it does not systematically report on the results of these measures. State officials told us that if they had more performance-based information, they could more easily present justification for funding that the United States directs to IAEA by providing Congress more information about the results and impact of U.S. contributions to the program. In 2005, we reported that IAEA did not systematically measure the results of its nuclear security activities and recommended that State work with IAEA and its member countries through the agency's Board of Governors to consider developing such measures.[42] Since that time, IAEA has developed the following two performance measures for the agency's nuclear security efforts:
    - number of countries that have established or improved national nuclear security systems on the basis of advice or assistance from the agency, and
    - number of countries implementing national nuclear security systems on the basis of agency assistance.

The Director of IAEA's Office of Nuclear Security told us that the agency is using these measures to assess the program's performance. However, IAEA has not systematically reported on the results of these performance measures. For example, in the agency's annual report for 2011, with regard to its nuclear security program, IAEA reported on the number of IPPAS missions undertaken annually and the number of training events facilitated related to nuclear security. However, IAEA did not report any results related to the two performance measures outlined in its program and budget report, and the Director of the Office of Nuclear Security could not provide any additional information documenting the results of the performance measures or how the agency used them to assess its progress in achieving the nuclear security program's goals. In our 2005 report, we stated that IAEA officials said that it is difficult for the agency to link nuclear security efforts, such as recommendations stemming from advisory missions, to results, such as whether recommendations were implemented.[43] During this review, IAEA officials reaffirmed this point and attributed this difficulty to the agency's limited advisory role, noting that countries are not obligated to follow its guidelines or implement its recommendations. Nonetheless, reporting on performance measures is important to ensure that IAEA can enhance the international community's knowledge of the impact and effectiveness of the nuclear security program. Because IAEA has not systematically reported on the results of its performance measures, member countries and the international community at large cannot gauge the extent to which IAEA is meeting its goals or the nuclear security program's impact and effectiveness.

IAEA Is Making Progress in Implementing a Nuclear Fuel Bank, but Uncertainties Exist Concerning Its Value, Potential Market Impact, and Long-term Operation

IAEA is making progress in establishing an international nuclear fuel bank containing LEU by 2014, but several experts and foreign officials told us they were uncertain about the fuel bank's value, potential market impact, and long-term operation. Further, IAEA's fuel bank is one of several guaranteed fuel supply options currently planned or in operation worldwide—including one in the United States—which raises questions about possible duplication of efforts.

Nuclear Fuel Bank Is Expected to become Operational by 2014

Since IAEA's Board of Governors approved in December 2010 the establishment of a nuclear fuel bank to provide a guaranteed fuel supply, the agency has made progress in its plans to operationalize the bank by 2014. IAEA and U.S. officials told us they expect the bank to have two benefits: (1) to serve as an assured source of nuclear fuel for countries with civilian nuclear power programs in the event of a supply disruption and (2) to offer countries an alternative to the development of sensitive nuclear fuel enrichment technologies, as such technologies could also be used to produce nuclear material—including highly enriched uranium (HEU)—for nuclear weapons.

IAEA officials responsible for the fuel bank told us that, since April 2012, the fuel bank project has shifted from the design to the implementation phase. During this phase, the agency plans to (1) negotiate a host state agreement with Kazakhstan concerning, among other things, the physical security, legal status, and safety of the bank,[44] (2) purchase approximately 60 metric tons of LEU for fuel, and (3) manage the bank's budget. As part of this process, the agency is preparing a site assessment in Kazakhstan that includes a review of the proposed site— the Ulba Metallurgical Plant, located in the city of Ust-Kamenogorsk—which manufactured nuclear fuel for the Soviet Union. Figure 4 shows a map of Kazakhstan that includes the location of the proposed site at Ust-Kamenogorsk. IAEA has also moved oversight of the fuel bank from its Office of Nuclear Energy to the Director General's Office for Policy. According to NNSA officials, this transfer was done to ensure that the project receives the highest level of attention by the agency and also to demonstrate its importance to IAEA's member countries.

According to a statement issued by IAEA's Director General in September 2012, the agency is reviewing the necessary legislative and regulatory framework that is required to ensure that the bank will operate consistently with IAEA safety standards, security guidelines, and safeguards requirements. The statement also noted that the results of these reviews would be incorporated in early 2013 into a plan that will make the bank fully operational. In addition, IAEA officials also told us that the site assessment would indicate whether physical security upgrades are required to accommodate the proposed fuel bank.

According to IAEA officials, the agency's Board of Governors has developed criteria for countries that could purchase LEU fuel from the bank. To be eligible to purchase fuel, a country has to meet the following criteria: (1) the country is experiencing a supply disruption of fuel for a nuclear power plant and is unable to obtain fuel by other means, (2) IAEA has concluded that there has been no diversion of declared nuclear material from the country and that there are no issues related to the country's safeguards implementation that are under consideration by the Board of Governors, and (3) the country has a comprehensive safeguards agreement in force. In addition, the country receiving fuel from the bank has to conclude a supply agreement with IAEA that contains the following additional criteria:

- The fuel can only be used for fuel fabrication for energy generation at a nuclear power plant;
- The fuel may not be used to manufacture any nuclear weapon or nuclear explosive device, or for any other military purpose;
- The country cannot further enrich, reprocess, retransfer, or re-export the fuel unless IAEA permits it to do so;
- The fuel is subject to applicable IAEA safeguards, safety standards, and physical protection measures; and
- The country must take responsibility for all liability for any nuclear damage that may be caused by a nuclear incident associated with the use, handling, storage, or transport of the fuel.

Sources: Department of State and GAO.

Figure 4. Map of Kazakhstan Showing the Proposed IAEA Nuclear Fuel Bank Site at Ust-Kamenogorsk.

Uncertainties Exist about the Fuel Bank's Value, Potential Market Impact, and Long-term Operation

We identified several uncertainties about the international nuclear fuel bank's value, potential impact on the international nuclear fuel market, and long-term operation, based on discussions with several experts and officials from State, NNSA, IAEA, and foreign missions. Specifically:

- *The fuel bank may never be used, raising questions about its value.* Officials from State, NNSA, and foreign missions, as well as some experts, told us that the fuel bank may never be used. For example, a senior State official told us that there are currently numerous countries that are willing and stable suppliers of LEU fuel; therefore, a country that needs fuel for civilian nuclear power plants already has willing suppliers from whom they can purchase it. NNSA officials told us that IAEA's fuel bank is intended to be used only as a last resort and that it will likely never be used. In addition, officials from some foreign missions told us that they were uncertain about the bank's true purpose, given that, in their view, there is a ready supply of LEU fuel worldwide and numerous suppliers. Officials from two of these foreign missions also described the bank as "a solution looking for a problem." Notwithstanding these concerns, some experts and foreign mission representatives who support the fuel bank told us that a better measure of the bank's success is in providing countries an assurance of supply, and one expert described the bank as a crisis avoidance tool.

  In addition, officials from one foreign mission told us that the countries the bank is targeting—those that are considered a nuclear proliferation risk, such as Iran—are suspicious of the bank's true purpose. As a result, those countries of greatest proliferation concern are not likely to use the bank under any circumstances.

Furthermore, according to State officials, some countries, such as Iran, are not eligible to participate due to past safeguards violations. Therefore, as a former Head of IAEA's Department of Safeguards told us, the bank does not solve the problem of Iran's continued nuclear fuel enrichment activities. Moreover, officials from several foreign missions, including some who support the bank, told us that certain developing countries that are members of the Non-Aligned Movement and the Group of 77[45] are concerned that the bank might be used to constrain, or at least discourage, countries' rights to pursue peaceful uses of nuclear energy under Article IV of the NPT by developing their own enrichment facilities.[46] Some experts we interviewed also stated that countries within the Non-Aligned Movement view the bank as IAEA's attempt to deprive them of their rights of peaceful use of nuclear energy. Officials from a foreign mission that supports the bank told us that this concern still exists, despite the Board of Governors' vote to approve the bank in 2010.[47]

- *The bank's potential impact on the international nuclear fuel market is uncertain.* The conditions that IAEA's Board of Governors has placed on the fuel bank include a stipulation that the use of the bank will not affect the international nuclear fuel market. IAEA plans to purchase 60 metric tons of LEU to supply the bank in advance of its startup. According to NNSA officials, the greatest challenge is identifying the point at which IAEA's fuel bank moves from serving as a safety net for countries in need of fuel to becoming a disruption to the nuclear fuel market. Similarly, a senior State official told us that some fuel supplier countries have expressed concern that the bank could disrupt the market and affect fuel prices. While IAEA has not released an assessment of the fuel bank's potential market impact, other studies indicate that a new source of supply could impact the international fuel market. For example, a 2010 study by the Brookings Institution queried nuclear industry officials on the feasibility of multilateral nuclear approaches, including an IAEA-administered nuclear fuel bank. According to the study, commercial nuclear fuel vendors are concerned about the implications of a fuel bank on the current market system that has, to date, had no supply issues. The study also reported that nuclear industry officials are worried that a fuel bank would artificially force down prices to levels that are uncompetitive for private vendors. In addition, according to a 2006 report by the Pacific Northwest National Laboratory,[48] the long-term impact of having assured fuel on the international nuclear market is unclear, including the effect that an assured fuel supply could have on the competitiveness of U.S. commercial fuel. For example, an influx of LEU on the market could affect market structure, potentially pushing a marginal supplier of uranium or enrichment out of the market, discouraging an existing supplier from making a necessary investment to upgrade its capacity, or depressing investment in mining and other operations. The report also stated that there has been significant discussion between industry and government regarding how much LEU can be introduced into the nuclear fuel market without causing a serious market disruption and added that a wide variety of variables can impact the market.

  IAEA officials who are responsible for managing the fuel bank told us that the bank will not affect the international nuclear fuel market for two reasons: (1) the agency will replenish the bank's inventory by procuring one reload of LEU for every reload of fuel that it supplies, making the bank market-neutral; and (2) the agency approved an LEU procurement plan in September 2011 that specifically addresses the avoidance of market impact during the initial inventory acquisition for the fuel bank. However, we have not reviewed the plan and its assessment of the bank's potential impact on the market, as the plan is confidential and not publicly available. As a result, it is impossible to evaluate whether IAEA's fuel bank could adversely affect the international market. Nonetheless, IAEA officials also told us that, in their view, the 60 metric tons of LEU the agency is planning to purchase is not enough to have a significant impact on the market worldwide.

- *IAEA does not have a plan in place for the long-term operation and funding of the fuel bank.* The nuclear fuel bank is the first project of its kind that IAEA has ever undertaken, is large in scope, and requires complex planning and coordination across different departments within the agency. While IAEA has given countries briefings about the fuel bank's status, the agency has not yet developed a long-term plan that addresses key aspects of the project, although agency officials responsible for managing the bank told us that they intend to provide a detailed financial plan to the Board of Governors in 2013. Among other things, they stated that this plan would identify the long-term funding needs for the bank. We have previously reported that a long-term funding plan should include effective strategies to help set priorities and allocate resources, including staffing, to inform decision-making and help ensure accountability.[49] Such priority-setting and resource allocation is especially important in a fiscally constrained environment, such as the one that IAEA faces, and should include information on the levels and sources of funding needed for the ongoing maintenance and operation of the fuel bank. IAEA officials told us that funding for the bank will be kept to a minimum and come solely from extra-budgetary contributions, with sufficient resources set aside to cover operation of the bank for a period of time to be determined by the Director General. The agency's Board of Governors agreed to rely on extra-budgetary funds because of the zero-real-growth budget policy under which the agency generally operates and the resistance among countries to increase the agency's regular budget to fund the project. As stated earlier, IAEA has received $159.4 million in extra-budgetary cash contributions and pledges from the United States and other donors toward the establishment of the fuel bank. However, without a long-term plan in place, it is unclear whether IAEA will need funding beyond the initial $150 million and, if so, how the agency would be able to financially support the bank's ongoing maintenance and operation.

IAEA's Nuclear Fuel Bank Is One of Several Guaranteed Fuel Supply Options Available

IAEA's nuclear fuel bank is one of several guaranteed, multilateral fuel supply options that are currently available to countries seeking an assured LEU fuel source in the event of a supply disruption. Several countries, including the United States, have independently established a fuel bank or similar mechanism to provide eligible countries with a guaranteed fuel supply. The number of fuel banks raises questions about a potential duplication of effort among these similar options.

Several Guaranteed Fuel Supply Options Are Available

We identified three options available for countries seeking an assured supply of LEU fuel in the event of a supply disruption, in addition to IAEA's planned nuclear fuel bank. These efforts have independently been undertaken by the United States, Russia, and the United Kingdom. Each of these initiatives is described below.

- *United States*. In addition to supporting IAEA's nuclear fuel bank, the United States has created its own guaranteed fuel supply option, called the American Assured Fuel Supply, to serve as a backup fuel supply to eligible foreign or domestic recipients. In 2005, the Department of Energy (DOE) announced that 17.4 metric tons of U.S. surplus HEU from a former nuclear weapons plant would be downblended to LEU to be used as a U.S. fuel bank. NNSA officials told us that the U.S. fuel bank has been operational since 2011, although it had not been used as of February 2013. According to an announcement in the *Federal Register* on August 18, 2011, the purpose of the U.S. fuel bank is to provide an assured supply of LEU fuel in the event of a supply disruption where all other market options have been exhausted. This fuel bank is also intended to help countries pursue nuclear power without the burden of producing their own fuel, while curbing the spread of sensitive technology. To be eligible to purchase the fuel, recipients, through their governments, must meet certain nonproliferation criteria.

  NNSA has made little information publicly available about the details of the U.S. fuel bank, apart from the *Federal Register* announcement and an accompanying press release. For example, NNSA has not released key information about the U.S. fuel bank's operations, such as whether the bank is being used, estimated demand, potential impact on the international nuclear fuel market, or any controls to mitigate such impacts. In May 2012, DOE issued a Secretarial Determination that stated, among other things, that certain planned DOE sales and transfers of excess uranium—including the transfer of 60 metric tons of LEU to a contractor in creating the U.S. fuel bank— would have no adverse material impact on the U.S. uranium mining, conversion, or enrichment industries.[50] NNSA officials acknowledged that the analysis did not cover the potential impact of the U.S. fuel bank's 230 metric tons of LEU on the international nuclear fuel market. As we noted in reference to the IAEA nuclear fuel bank, some studies indicate that a new source of LEU fuel could affect the international fuel market, including the competitiveness of U.S. commercial fuel. NNSA officials told us that there were no requirements for the agency to conduct any studies that would provide this type of information. Nonetheless, without more information, the extent to which the U.S. fuel bank may be used, its potential impact on the international nuclear fuel market, or any controls to mitigate such impact is unclear.

  In addition, during our review, we obtained some key information from NNSA officials about the U.S. fuel bank's financial status that had not been previously reported. For example, NNSA officials told us that as of December 2012, Nuclear Fuel Services, a contractor in Erwin, Tennessee, had completed downblending 17.4 metric tons of HEU, yielding 290 metric tons of LEU. Of that amount, approximately 230 metric tons of LEU have been made available for the U.S. fuel bank, which is stored at a fuel fabrication facility in Columbia, South Carolina, while the remaining 60 metric tons were used to compensate the contractor through what NNSA describes as a barter arrangement.[51] The value of this arrangement is approximately $145.3 million, composed of (1) $105.3 million to cover the contractor's downblending services and (2) approximately $40 million to cover the cost of diluent (natural uranium that is blended with HEU to produce LEU). As compensation for its storage of the LEU for the U.S. fuel bank, the contractor is also allowed to borrow a portion of the LEU for its ongoing business activities, provided the contractor meets certain conditions, including replacing any LEU it uses. The storage portion of the contract extends for 10 years, through June 2017, but there are also options to extend the contract in 5-year increments. However, NNSA has not published this information in publicly available documents, such as DOE's *Excess Uranium Inventory Management Plan*.[52]

In addition to NNSA's $145.3 million contract, NNSA officials told us that the agency spent a total of $28.3 million in fiscal years 2002 through 2012 to prepare the HEU for downblending and to support the project. Specifically, NNSA officials said these funds were used to retrieve the HEU from NNSA's storage repository, perform inspections, sample, process, pack, and ship the HEU to the contractor and, in later years, to provide program management support such as quarterly contract reviews. However, NNSA has not quantified the specific costs per year associated with this aspect of creating the U.S. fuel bank in its annual budget justification reports to Congress or in other publicly available documents. We have previously reported that Federal Accounting Standards Advisory Board guidance calls for the identification of the full cost of federal programs and activities, and congressional requesters have emphasized that NNSA should establish budgets that reflect total program costs and that these budgets should be more transparent to oversight.[53] Moreover, DOE's 2008 *Excess Uranium Inventory Management Plan* states that the agency will manage its excess uranium inventory through controlled sales or transfers in a transparent manner that will result in the U.S. government's receipt of reasonable value for any such transactions.

- *Russia.* In 2009, IAEA's Board of Governors approved a Russian-established fuel bank housed in Angarsk, Russia, that operates under the auspices of IAEA. Specifically, according to an IAEA report, a country facing a fuel supply cutoff would apply to IAEA to access the fuel bank, and IAEA's Director General would then assess whether the applicant country meets the criteria for access. Similar to IAEA's nuclear fuel bank, fuel from the Russian bank is available as a backup supply to any eligible IAEA member country that faces a noncommercial or technical disruption of fuel supply for its civilian nuclear power reactors. According to IAEA officials, this bank is funded entirely by Russia and uses Russian fuel, whereas IAEA's fuel bank is being funded by the $150 million in extra-budgetary contributions from certain member countries and other donors.
- *United Kingdom.* In 2011, IAEA's Board of Governors approved a proposal by the United Kingdom to create a nuclear fuel assurance program that would ensure the continued supply of nuclear fuel in the event a country's supply is disrupted for political reasons. Unlike the IAEA, Russian, and U.S. fuel banks, this guaranteed fuel supply option does not call for a physical stockpile of fuel. Instead, it is intended to

provide a contractual agreement between IAEA, a supplier country, and a recipient country to guarantee an uninterrupted supply of fuel. According to IAEA, the agency's status as a co-signatory would play a central role in building confidence in such contracts.

Multiple Fuel Supply Options May Result in Duplication of Effort

We identified potential overlapping functions among the United States' nuclear fuel bank and these other guaranteed fuel supply options, as they are all similar in purpose and intent.[54] Although some of these options are better developed than others, they all have a common goal. Specifically, each one seeks to provide a guaranteed supply of nuclear fuel to eligible countries to reduce the risk of nuclear proliferation. Consequently, they all appear to be targeting a similar set of countries: those countries that are looking to develop or sustain civilian nuclear power programs without having to develop their own nuclear fuel enrichment capabilities. However, it is unclear how the existence of multiple guaranteed fuel supply options will directly reduce potential proliferation risks because those countries of greatest proliferation concern may not be likely—or eligible—to use these fuel supply options.

Notably, there is potential duplication between the U.S. nuclear fuel bank, which already exists, and the IAEA fuel bank, which is expected to begin operation by 2014. There may be significant financial implications associated with the United States' supporting two nuclear fuel banks. Specifically, the United States has already expended resources with a total value of approximately $223.1 million on these two fuel banks through fiscal year 2012: $49.5 million for the IAEA bank and about $173.6 million for the U.S. bank. Given the fiscal pressures that the United States currently faces, it is important for NNSA to review and assess the need for a U.S. fuel bank and report on the results of such an assessment. As we have reported in the past,[55] consolidating programs sharing common goals and implementing similar projects can maximize limited resources and may achieve potential cost savings or other programmatic and administrative efficiencies. In addition, we have reported that conducting an assessment to better understand the extent to which programs may overlap with each other is important to mitigate the risk of unnecessary duplication. However, as of February 2013, NNSA officials told us they had not formally assessed the extent to which the U.S. fuel bank may overlap with IAEA's fuel bank. As a result, NNSA may lack information on the extent to which unnecessary duplication between the U.S. and IAEA fuel banks may exist.

NNSA and IAEA officials told us that having several backup fuel supply options increases the availability of LEU fuel worldwide to serve as a safety net in the event of a supply disruption or shortage. Thus, according to IAEA officials, the existence of multiple fuel assurance programs would help ensure that a country interested in developing or expanding its civilian nuclear power program would have ready access to fuel and not be dependent on one supplier but could choose from a number of fuel supply options. Furthermore, IAEA officials told us that the greater the number of these fuel supply options, the lesser the likelihood that a country will develop its own enrichment capability, thus reducing potential nuclear proliferation risks. These officials also stated that some element of duplication is inherent in the concept of assurance but noted that there is a point at which multiple efforts have diminishing returns. Moreover, having multiple safety net suppliers decreases the likelihood that any one of these options will ever be used.

CONCLUSION

IAEA has continued to strengthen its safeguards program since we last reported on it in 2005. In particular, IAEA has successfully encouraged more countries to conclude an Additional Protocol and, as appropriate for certain countries, to amend or rescind small quantities protocols, or to bring comprehensive safeguards agreements into force. Despite this progress, IAEA continues to face certain challenges that may limit its ability to further enhance the effectiveness and efficiency of the safeguards program. First, the agency has not clearly defined and communicated how it will implement the state-level concept, and several countries have raised concerns as to whether IAEA would implement it in an objective, nondiscriminatory manner that does not exceed its authority to collect information about countries' nuclear activities. Without widespread acceptance among member countries, IAEA may not be able to implement this next phase in the evolution of safeguards in all countries with safeguards agreements by 2014. Second, the agency has not identified the human capital, technological, or other resources it needs to broaden implementation of the state-level concept—as well as the associated long-term costs and benefits of such resources. Without assessing and quantifying the long-term resource needs and the associated costs and benefits, it is unclear what expertise, technology, or other resource gaps may exist.

In addition, while IAEA has increased training and other efforts to build countries' capacity to improve the physical security of their nuclear material and facilities, the agency continues to rely heavily on extra-budgetary contributions, which vary from year to year and are often designated by donors for specific projects in certain countries. As a result, it is difficult for the agency to plan and implement projects where security needs may be greatest. In this context, IAEA has not assessed its long-term resource needs for nuclear security activities and has not identified how it will meet those needs through the regular budget, extra-budgetary contributions, or both. Without more predictable long-term financial resources, the agency cannot adequately plan ahead and ensure that the greatest security needs are addressed in a timely manner. Further, while IAEA has developed performance measures for its nuclear security program since our 2005 report, the agency does not systematically report on the results of these performance measures. Without systematically reporting on these results, member countries and the international community at large cannot gauge the extent to which IAEA is achieving its goals or the nuclear security program's impact and effectiveness.

The concept of an international nuclear fuel bank as a nonproliferation tool may have merit, but uncertainties exist about its value, potential market impact, and long-term operation. In particular, it is unclear what value the IAEA fuel bank adds to nuclear nonproliferation efforts worldwide. Furthermore, it is impossible to determine whether IAEA's fuel bank would interfere with the operation of the international nuclear fuel market because the agency has not made publicly available its assessment of the bank's potential market impact. Two years after IAEA's Board of Governors voted to approve the nuclear fuel bank, the agency also has not yet developed a plan for the bank's long-term operation and funding. Without a long-term plan that includes strategies to help set priorities and allocate resources to guide the bank beyond its first few years, the long-term prospects for the fuel bank's viability are uncertain. In the event that IAEA needs funds beyond the initial $150 million, the United States—the largest single-country donor to the fuel bank—and other donors will likely expect additional information about the bank's long-term operation before making further contributions.

In addition to supporting IAEA's nuclear fuel bank, the United States has created its own fuel bank to serve as a safety net fuel supply to eligible countries. However, it is unclear how the existence of multiple guaranteed fuel supply options will directly reduce potential proliferation risks because those countries of greatest proliferation concern may not be likely—or eligible—to use the U.S. fuel bank. Furthermore, NNSA has released little information about the U.S. bank's operations, such as whether the bank is being used, estimated demand, potential impact on the international market, or any controls to mitigate the potential market impact. Moreover, NNSA has not released information on the financial status of the U.S. fuel bank—including a 10-year contract that transferred government assets worth approximately $145.3 million, plus $28.3 million in appropriated funds from fiscal years 2002 through 2012—in NNSA's annual budget justification reports or other publicly available documents. Finally, given the overlapping purposes of IAEA's and the United States' fuel banks, there may be duplication of effort that could have significant financial implications. In light of the fiscal pressures the United States faces, and the United States' financial support of IAEA's fuel bank, it is important for NNSA to review and assess the need for a U.S. fuel bank and report on the results of such an assessment.

RECOMMENDATIONS FOR EXECUTIVE ACTION

We are making nine recommendations in this report.

To ensure that safeguards and nuclear security resources are allocated and spent in the most effective and efficient manner, the Secretary of State should work with IAEA and its member countries through the agency's Board of Governors to consider the following four actions:

- clearly define and communicate how the agency will implement the state-level concept in an objective, nondiscriminatory manner that does not exceed IAEA's authority to collect information about countries' nuclear activities;
- conduct an assessment of the human capital, technological, and other resources the agency may need to broaden implementation of the state-level concept—and the associated long-term benefits and costs;
- evaluate the nuclear security program's long-term resource needs and assess whether the agency's heavy reliance on extra-budgetary contributions is sufficient to plan and meet those needs; and
- systematically report on the results of the agency's performance measures for the nuclear security program to allow member countries and the international community at large to gauge the extent to which the agency is achieving its goals or the nuclear security program's impact and effectiveness.

To ensure that IAEA's nuclear fuel bank is implemented in the most effective and efficient manner, the Secretary of State should work with IAEA and its member countries through the agency's Board of Governors to consider the following two actions:

- prepare a publicly available assessment of the potential impact of IAEA's nuclear fuel bank on the international nuclear fuel market; and
- ensure that the agency's plan for the long-term operation and funding of the nuclear fuel bank includes strategies to help set priorities and allocate resources for the viability of the fuel bank that include, for example, information on the levels and sources of funding needed for the ongoing maintenance and operation of the fuel bank.

To ensure the most effective, efficient, and transparent use of U.S. government resources, we recommend that the Administrator of NNSA take the following three actions:

- review and assess the need for the U.S. nuclear fuel bank, based on the United States' financial support of IAEA's fuel bank and the potential for duplication of efforts, and report on the results of this assessment;
- report key information on the U.S. nuclear fuel bank's operations, such as its use to date, estimated demand, potential impact on the international nuclear fuel market, or any controls to mitigate the potential market impact; and
- review and report on the financial status of the U.S. nuclear fuel bank, including its costs to date and any ongoing costs related to the bank, in NNSA's annual budget justification reports or other publicly available documents.

AGENCY COMMENTS AND OUR EVALUATION

We provided a draft of this report to State, NNSA, the Nuclear Regulatory Commission, and IAEA for review and comment. In its written comments, State noted that our report provides timely and useful information on several aspects of IAEA's continuing work in the vital areas of safeguards and nuclear security. In addition, State commented that our report accurately recognizes that continued and focused work in the areas identified will require the United States to work closely with IAEA and other member countries to achieve success. Of the six recommendations directed to it, State agreed with three recommendations, disagreed with two recommendations, and did not comment on our recommendation that it should work with IAEA to consider systematically reporting on the results of the agency's performance measures for the nuclear security program. NNSA agreed with all three recommendations directed to it. The Nuclear Regulatory Commission and IAEA did not provide written comments. All three U.S. agencies, as well as IAEA, provided technical comments on our draft report, which we incorporated as appropriate.

In its written comments, State agreed with our recommendations that it should work with IAEA and other member countries to consider the following three actions: (1) clearly define and communicate how IAEA will implement the state-level concept in an objective, nondiscriminatory manner that does not exceed its authority to collect information about a country's nuclear activities; (2) conduct an assessment of the human capital, technological, and other resources the agency may need to broaden implementation of the state-level concept—and the associated long-term benefits and costs; and (3) ensure that the agency's plan for the long-term operation and funding of IAEA's nuclear fuel bank includes strategies to help set priorities and allocate resources for the viability of the fuel bank, such as information on the levels and sources of funding needed for the ongoing maintenance and operation of the fuel bank.

However, State disagreed that it should work with IAEA and other member countries to consider the following two actions: (1) evaluate the nuclear security program's long-term resource needs and whether the agency's heavy reliance on extra-budgetary contributions is sufficient to plan and meet those needs, and (2) prepare a publicly available assessment of the potential impact of IAEA's nuclear fuel bank on the international nuclear fuel market. Specifically, State commented that it believes a resource assessment or long-term planning for the nuclear security program is not possible because IAEA assists member countries by responding to specific requests that cannot be anticipated and then addresses those requests based on staffing and resource availability. State also noted that it does not believe IAEA should spend resources assessing whether extra-budgetary funding is sufficient because such funding is, by nature, voluntary, unpredictable, and often comes with conditions. In addition, State indicated that IAEA will continue to rely heavily on extra-budgetary contributions for nuclear security, given the limited regular budget for the program.

Notwithstanding State's comments, we continue to believe that long-term strategic and budgetery planning is important to identify priorities and needs and that the absence of such information hinders IAEA's ability to ensure that resources are directed to the greatest nuclear security needs. Given IAEA's extensive knowledge of nuclear security issues worldwide— and after managing the program for more than 10 years—we believe it is reasonable to expect the agency to forecast, prioritize, and plan beyond the agency's 2-year budget cycle. While IAEA has started to develop country-specific nuclear security support plans to identify individual countries' needs, the agency does not have an overall plan that looks comprehensively at nuclear security needs globally and identifies long-term priorities and resource needs for the agency. In our view, such information would add a greater degree of transparency to IAEA's nuclear security program and provide member countries with a clearer understanding of how they might assist the agency in achieving its long-term goals while maximizing limited resources. We agree with State that IAEA will likely continue to rely heavily on extra-budgetary contributions for nuclear security, which, as we have reported, exceeded 80 percent of IAEA's total nuclear security funding both during our prior review in 2005 and our current review. However, this may not be a sustainable long-term funding plan, and the agency may be missing opportunities to justify a potential increase to its limited regular budget and, therefore, to rely less on extra-budgetary contributions. As noted in our report, such extra-budgetary contributions typically fluctuate from year to year and may be designated by donors for projects that are not the agency's highest priorities. We recognize that IAEA currently operates under a zero-realgrowth budget environment, but given the fiscal constraints faced by major donors, including the United States, IAEA cannot necessarily assume that donors will continue to make extra-budgetary contributions at the same levels as in the past.

Regarding our recommendation that IAEA should consider preparing a publicly available assessment of the potential impact of IAEA's nuclear fuel bank on the international nuclear fuel market, State indicated that such an assessment "would be a questionable use of the Agency's resources." We disagree with State's position because, as noted in our report, IAEA officials told us that they have already conducted such an assessment but have so far kept the assessment confidential. The point of our recommendation is for IAEA to publicly release the existing assessment so that member countries and others in the international community can evaluate it and draw their own conclusions. We recognize that IAEA's fuel bank is intended to be used only as a guaranteed supply option in the event of an unanticipated market failure. Nonetheless, some stakeholders have raised concerns, as noted in our report, about the potential market impact of such a multilateral nuclear fuel bank, regardless of its size. In fact, a senior State official told us that some fuel supplier countries have expressed concern that IAEA's planned fuel bank could disrupt the market and affect fuel prices. In our view, unless IAEA publicly releases its assessment, there will likely continue to be debate about whether IAEA's fuel bank could have unintended consequences. In addition, State commented that there is no danger the IAEA fuel bank would "flood" the international nuclear fuel market because the fuel would never be offered en masse or under a long-term supply contract. Our report does not suggest that IAEA's fuel bank would flood the market but, rather, that its potential effect on the market is unclear. As stated in our report, one of the IAEA Board of Governors' conditions for the bank is that its use will not affect the international nuclear fuel market. Thus we continue to believe that our recommendation for IAEA to publicly release its assessment of the fuel bank's potential market impact has merit.

In its written comments, NNSA agreed with all three recommendations directed to NNSA to help ensure the most effective, efficient, and transparent use of U.S. government resources in reference to the U.S. nuclear fuel bank. Specifically, our recommendations are for NNSA to take the following three actions: (1) review and assess the need for the U.S. nuclear fuel bank, based on the United States' financial support of IAEA's fuel bank and the potential for duplication of efforts, and report on the results of this assessment; (2) report key information on the U.S. nuclear fuel bank's operations, such as its use to date, estimated demand, potential impact on the international nuclear fuel market, or any controls to mitigate the potential market impact; and (3) review and report on the financial status of the U.S. nuclear fuel bank, including its costs to date and any ongoing costs related to the bank, in NNSA's annual budget justification reports or other publicly available documents. Regarding the third recommendation, NNSA commented that "there are no future costs anticipated to NNSA, as the ongoing compensation for the storage of the [LEU] for the fuel bank is covered by barter arrangement whereby the contractor can use a portion of the LEU as working inventory." Based on this comment and additional technical information provided by NNSA, we have added clarifying language in our report. Specifically, we added that the contractor is allowed to borrow a portion of the LEU fuel for its ongoing business activities, provided the contractor meets certain conditions, including replacing any LEU it uses. In addition, we noted that the storage portion of the contract extends for 10 years, through June 2017, and that there are options for extending the contract in 5-year increments. However, until NNSA extends the contract or potentially awards it to another party, it is premature to speculate, in our view, as to whether NNSA may incur additional storage or other costs related to the U.S. fuel bank in the future.

David C. Trimble
Director, Natural Resources and Environment

APPENDIX I. OBJECTIVES, SCOPE, AND METHODOLOGY

The objectives of our review were to examine: (1) changes that have occurred since 2005 and challenges, if any, that the International Atomic Energy Agency (IAEA) faces in carrying out its safeguards program; (2)changes that have occurred since 2005 and limitations, if any, regarding IAEA's nuclear security program; and (3) the status of IAEA's planned international nuclear fuel bank.

For all three objectives, we traveled to Vienna, Austria, in June 2012 andspoke with IAEA officials responsible for implementing the safeguards,nuclear security, and fuel bank programs. While in Vienna, we also conducted interviews with a nonprobability sample[56] of officials from 15 countries, in addition to the U.S. Mission to the United Nations SystemOrganizations in Vienna (U.S. Mission), to obtain their views on these programs.

To select these countries, we identified criteria and selected countries that met at least three of the criteria we identified. The criteria for selection included IAEA member countries that:

- possess nuclear weapons,
- reprocess nuclear material,
- are members of the IAEA Board of Governors for 2011-2012,
- are members of the Non-Aligned Movement,
- have nuclear power or plan to have nuclear power reactors,
- brought an Additional Protocol into force, and
- consume the greatest share of IAEA safeguards resources.

Given the aforementioned criteria, we identified and invited 25 geographically diverse countries to participate in interviews. Of the 25 countries, 15 responded to our invitation. Therefore, we completed in-person interviews with officials from the following countries: Brazil,Canada, Chile, China, France, Germany, India, Israel, Japan, Pakistan, Russia, South Korea, Turkey, the United Arab Emirates, and the UnitedKingdom. Throughout this report, we distinguish between statements made by the U.S. Mission and other countries (foreign missions), and we use the term "some" when referring to two or three countries and "several" when referring to four or more countries.

In addition, we interviewed 15 experts who specialize in nuclear nonproliferation issues and are knowledgeable about IAEA's safeguards, nuclear security, and/or fuel bank programs. Specifically, we met with 11 experts from 9 different academic think tanks and nongovernmental organizations—such as the Belfer Center for Science and International Affairs at Harvard University's John F. Kennedy School of Government, the World Institute for Nuclear Security, and the Nuclear Threat Initiative—and 4 former senior IAEA officials. The four former senior IAEA officials consisted of the immediate past two Heads of the Department of Safeguards, the former Director of the Office of Nuclear Security, and the prior Head of the Department of Management. To select the experts we interviewed, we identified and prepared a list of nuclear nonproliferation experts based on several factors, including the following:

- their direct knowledge of IAEA from their recent prior work at the agency regarding the safeguards, nuclear security, and/or fuel bank programs;
- their extensive knowledge of IAEA's safeguards, nuclear security, and/or fuel bank programs, based on their work at academic institutions or nongovernmental organizations;
- their prior experience in meeting with GAO as subject matter experts; and
- referrals from other nuclear nonproliferation experts (snowball sampling).

Throughout this report, we use the term "some" when referring to two or three experts and "several" when referring to four or more experts.

For all three objectives, we also collected documents from, and interviewed, officials from the Department of Energy's (DOE) National Nuclear Security Administration (NNSA) and DOE national laboratories, including Brookhaven National Laboratory in New York, which manages U.S. technical assistance to IAEA's safeguards program. Further, we obtained the views of officials from the U.S. Mission on the changes and any challenges they identified for the IAEA safeguards and nuclear security programs since we last reported on them in 2005, as well as their views on IAEA's planned fuel bank.[57] We also met with officials from the Department of State's (State) Bureau of International Organization Affairs, Bureau of International Security and Nonproliferation, and Office of Multilateral Nuclear and Security Affairs in Washington, D.C., as well as officials from State who were knowledgeable about U.S. extra-budgetary contributions for each of the programs.

To examine the changes that have occurred since 2005 and challenges, if any, that IAEA faces in carrying out its safeguards program, we obtained and analyzed agency documentation related to the safeguards program, including the agency's biennial program and budget documents, annual reports, and long- and medium-term strategic plans. In June 2012, we met with IAEA officials from the Department of Safeguards, including the directors and staff responsible for implementing the state-level concept in Vienna, Austria. We also toured IAEA's Safeguards Analytical Laboratories in Seibersdorf, Austria, where environmental and nuclear material samples are analyzed. We also interviewed officials from DOE's Lawrence Livermore National Laboratory in California, which, as part of IAEA's international network of safeguards analytical laboratories, analyzes nuclear material samples collected by IAEA inspectors. Furthermore, we interviewed officials from the Nuclear Regulatory Commission and the Department of Defense, including a senior official from the U.S. Air Force Technical Applications Center, which operates one of the laboratories that conducts environmental sample analysis as part of IAEA's international network of safeguards analytical laboratories. In addition, we attended a workshop in May 2012 on "evolving the state-level concept" because it was dedicated to discussing IAEA's plans for implementing the state-level concept for safeguards and because it included attendees from IAEA and U.S. government agencies. The workshop was hosted by the Institute of Nuclear Materials Management and held at the University of Virginia in Charlottesville, Virginia. To identify IAEA's total regular budget and extra-budgetary cash contributions for the safeguards program in 2011, we indentified such information in the agency's annual report for 2011 and gathered and reviewed data from IAEA officials. To identify the United States' extra-budgetary cash contributions to the safeguards program in 2011, we gathered and reviewed data from State.

To examine the changes that have occurred since 2005 and limitations, if any, regarding IAEA's nuclear security program, we collected and analyzed documentation, including

IAEA's yearly reports to its Board of Governors on its nuclear security program, as well as IAEA's *Nuclear Security Plan 2010-2013*. We also met with officials from IAEA's Office of Nuclear Security and the Office of Legal Affairs in Vienna, Austria. Moreover, we met with NNSA officials to obtain their views on IAEA's security activities, budget, and management issues. In addition, we spoke with experts from nongovernmental organizations and academic think tanks, such as the Partnership for Global Security and Harvard University's Belfer Center, to obtain their perspectives regarding IAEA's nuclear security activities. To identify IAEA's total regular budget and extra-budgetary cash contributions for the nuclear security program in 2011, we reviewed such information in IAEA's annual report for 2011. To identify the United States' extra-budgetary cash contributions to the nuclear security program, we gathered and reviewed data from State.

To examine the status of IAEA's planned nuclear fuel bank, we spoke with officials in the Director General's Office for Policy in Vienna, Austria, to discuss IAEA's progress toward implementing the fuel bank. We also met with officials from the Nuclear Threat Initiative, which, alongside IAEA and the U.S. government, contributed funding for the fuel bank. Further, we met with NNSA officials to obtain additional information regarding IAEA's fuel bank, as well as the American Assured Fuel Supply (U.S. fuel bank). To identify IAEA's total budget on the fuel bank program, we obtained documentation and met with IAEA officials involved in the budgeting process. We also obtained U.S. budget data related to the U.S. fuel bank from NNSA officials, as well as gathered other information related to the U.S. fuel bank published in the *Federal Register*.

To assess the reliability of IAEA's budget data related to the safeguards, nuclear security, and fuel bank programs, we compared data from different published sources, including the *Federal Register*, and we met with IAEA and U.S. officials to discuss these data in detail. We also obtained and reviewed responses from key officials from State that addressed such areas as data entry, data access, quality control procedures, and data accuracy and completeness regarding the data they provided on U.S. extra-budgetary cash contributions for each program. Further, we analyzed other documentation, such as the external auditor's report on IAEA, as well as the agency's strategic planning documents. In addition, we obtained written responses from IAEA and U.S. officials to clarify discrepancies in the data we received. Throughout this report, we calculated all monetary figures for IAEA's safeguards, nuclear security, and fuel bank programs using the exchange rates published in the agency's annual report for 2011, which was the average exchange rate used by the United Nations in 2011 of $1.3893 to €1.00. Based on this work, we determined that data provided by IAEA to be sufficiently reliable for presenting the overall amount of money budgeted for each program in 2011. Additionally, we determined that the data on U.S. extra-budgetary cash contributions provided by State to be sufficiently reliable for presenting the overall amount of U.S. extra-budgetary cash contributions to IAEA in 2011. Finally, we determined that the data provided by NNSA and published in the *Federal Register* related to U.S. funding for IAEA's planned fuel bank to be sufficiently reliable for presenting the amount of funding that IAEA's fuel bank program has received to date from the United States.

We conducted this performance audit from January 2012 to April 2013 in accordance with generally accepted government auditing standards. Those standards require that we plan and perform the audit to obtain sufficient, appropriate evidence to provide a reasonable basis for our findings and conclusions based on our audit objectives. We believe that the evidence obtained provides a reasonable basis for our findings and conclusions based on our audit objectives.

APPENDIX II. STATUS OF COUNTRIES' SAFEGUARDS AGREEMENTS, ADDITIONAL PROTOCOLS, AND SMALL QUANTITIES PROTOCOLS IN FORCE WITH IAEA, AS OF DECEMBER 2012

Country	Comprehensive safeguards agreement[a]	Additional Protocol	Small quantities protocol
Non-nuclear weapon states			
Afghanistan	X	X	X
Albania	X	X	
Algeria	X		
Andorra	X	X	X
Angola	X	X	X
Antigua and Barbuda	X		X
Argentina	X		
Armenia	X	X	
Australia	X	X	
Austria	X	X	
Azerbaijan	X	X	X
Bahamas	X		X
Bahrain	X	X	X
Bangladesh	X	X	
Barbados	X		X
Belarus	X		
Belgium	X	X	
Belize	X		X
Benin			
Bhutan	X		X
Bolivia	X		X
Bosnia and Herzegovina	X		
Botswana	X	X	
Brazil	X		
Brunei Darussalam	X		X
Bulgaria	X	X	
Burkina Faso	X	X	X
Burundi	X	X	X
Cambodia	X		X
Cameroon	X		X
Canada	X	X	
Cape Verde			
Central African Republic	X	X	X

(Continued)

Country	Comprehensive safeguards agreement[a]	Additional Protocol	Small quantities protocol
Chad	X	X	X
Chile	X	X	
Colombia	X	X	
Comoros	X	X	X
Congo, Republic of the	X	X	X
Costa Rica	X	X	X
Côte d'Ivoire	X		
Croatia	X	X	X
Cuba	X	X	
Cyprus	X	X	
Czech Republic	X	X	
Democratic People's Republic of Korea (North Korea)[b]	X		
Democratic Republic of the Congo	X	X	
Denmark	X	X	
Djibouti			
Dominica	X		X
Dominican Republic	X	X	X
Ecuador	X	X	X
Egypt	X		
El Salvador	X	X	X
Equatorial Guinea			
Eritrea			
Estonia	X	X	
Ethiopia	X		X
Fiji	X	X	X
Finland	X	X	
Gabon	X	X	X
Gambia	X	X	X
Georgia	X	X	
Germany	X	X	
Ghana	X	X	
Greece	X	X	
Grenada	X		X
Guatemala	X	X	X
Guinea			
Guinea-Bissau			
Guyana	X		X
Haiti	X	X	X
Holy See	X	X	X

Country	Comprehensive safeguards agreement[a]	Additional Protocol	Small quantities protocol
Honduras	X		X
Hungary	X	X	
Iceland	X	X	X
Indonesia	X	X	
Iran, the Islamic Republic of	X		
Iraq	X	X	
Ireland	X	X	
Italy	X	X	
Jamaica	X	X	
Japan	X	X	
Jordan	X	X	X
Kazakhstan	X	X	
Kenya	X	X	X
Kiribati	X		X
Korea, Republic of (South Korea)	X	X	
Kuwait	X	X	X
Kyrgyzstan	X	X	X
Laos	X		X
Latvia	X	X	
Lebanon	X		X
Lesotho	X	X	X
Liberia			
Libya	X	X	
Liechtenstein	X		
Lithuania	X	X	
Luxembourg	X	X	
Macedonia, Republic of	X	X	X
Madagascar	X	X	X
Malawi	X	X	X
Malaysia	X		
Maldives	X		X
Mali	X	X	X
Malta	X	X	
Marshall Islands	X	X	
Mauritania	X	X	X
Mauritius	X	X	X
Mexico	X	X	
Micronesia			
Moldova, Republic of	X	X	X
Monaco	X	X	X
Mongolia	X	X	X

(Continued)

Country	Comprehensive safeguards agreement[a]	Additional Protocol	Small quantities protocol
Montenegro	X	X	X
Morocco	X	X	
Mozambique	X	X	X
Myanmar	X		X
Namibia	X	X	X
Nauru	X		X
Nepal	X		X
Netherlands	X	X	X
New Zealand	X	X	X
Nicaragua	X	X	X
Niger	X	X	
Nigeria	X	X	
Norway	X	X	
Oman	X		X
Palau	X	X	X
Panama	X	X	X
Papua New Guinea	X		X
Paraguay	X	X	X
Peru	X	X	
Philippines	X	X	
Poland	X	X	
Portugal	X	X	
Qatar	X		X
Romania	X	X	
Rwanda	X	X	X
St. Kitts and Nevis	X		X
St. Lucia	X		X
St. Vincent & the Grenadines	X		X
Samoa	X		X
San Marino	X		X
Sao Tomé and Principe			
Saudi Arabia	X		X
Senegal	X		X
Serbia	X		
Seychelles	X	X	X
Sierra Leone	X		X
Singapore	X	X	X
Slovakia	X	X	
Slovenia	X	X	
Solomon Islands	X		X

Country	Comprehensive safeguards agreement[a]	Additional Protocol	Small quantities protocol
Somalia			
South Africa	X	X	
Spain	X	X	
Sri Lanka	X		
Sudan	X		X
Suriname	X		X
Swaziland	X	X	X
Sweden	X	X	
Switzerland	X	X	
Syrian Arab Republic (Syria)	X		
Tajikistan	X	X	
Tanzania, United Republic of	X	X	X
Thailand	X		
Timor-Leste			
Togo	X	X	X
Tonga	X		X
Trinidad and Tobago	X		X
Tunisia	X		
Turkey	X	X	
Turkmenistan	X	X	
Tuvalu	X		X
Uganda	X	X	X
Ukraine	X	X	
United Arab Emirates	X	X	X
Uruguay	X	X	
Uzbekistan	X	X	
Vanuatu			
Venezuela	X		
Vietnam	X	X	
Yemen	X		X
Zambia	X		X
Zimbabwe	X		X
Nuclear weapon states[c]	**Voluntary offer safeguards agreement**	**Additional Protocol**	**Small quantities protocol**
China	X	X	
France	X	X	X[d]
Russian Federation (Russia)	X	X	
United Kingdom	X	X	X[d]
United States	X	X	X[d]

(Continued)

Non-NPT signatory countries[e]	Item-specific safeguards agreement	Additional Protocol	Small quantities protocol
India	X		
Israel	X		
Pakistan	X		

Source: GAO presentation of IAEA data.

[a] As noted earlier, IAEA does not officially recognize Taiwan as a state, but IAEA applies safeguards in Taiwan under two agreements that entered into force in October 1969 and December 1971. Taiwan has also implemented the measures of the Model Additional Protocol since October 1999.

[b] Although North Korea concluded a comprehensive safeguards agreement with IAEA in 1992, the agency has not been able to conduct all necessary safeguards activities provided for in the agreement since 1994 and, in January 2003, North Korea announced its withdrawal from the Treaty on the Nonproliferation of Nuclear Weapons (NPT).

[c] As of the end of December 2012, the five nuclear weapon states had safeguards agreements in force, called voluntary offer agreements, in which safeguards were implemented with regard to declared nuclear material in selected facilities in all five states. In addition, the five nuclear weapon states had Additional Protocols in force at that time.

[d] As of the end of December 2012, France, the United Kingdom, and the United States had operative small quantities protocols in reference to their respective territories in Latin America. These operative small quantities protocols are associated with these three countries' safeguards agreements pursuant to Additional Protocol I to the Treaty of Tlatelolco, which are different from their voluntary offer agreements under the NPT.

[e] As of the end of December 2012, these three non-NPT signatory countries had what are called item-specific safeguards agreements in force that require the application of safeguards to nuclear material, facilities, and other items specified in the relevant safeguards agreement.

End Notes

[1] IAEA is an autonomous international organization affiliated with the United Nations, established in Vienna, Austria, in 1957. The agency has the dual role of promoting the peaceful uses of nuclear energy by transferring nuclear science and technology through its nuclear science and applications and technical cooperation programs, and verifying, through its safeguards program, that nuclear material subject to safeguards is not diverted to nuclear weapons or other proscribed purposes.

[2] Under Article II of the NPT, each non-nuclear weapon state party agrees, among other things, not to receive any transfer whatsoever of nuclear weapons or other nuclear explosive devices; not to manufacture or otherwise acquire nuclear weapons or other nuclear explosive devices; and not to seek or receive any assistance in the manufacture of nuclear weapons or other nuclear explosive devices.

[3] Under Article III of the NPT, each non-nuclear weapon state party agrees, among other things, to accept IAEA safeguards on all source or special fissionable material in all peaceful nuclear activities within the territory of such state, under its jurisdiction, or carried out under its control anywhere.

[4] Under Article IV of the NPT, all parties undertake, among other things, to facilitate the exchange of peaceful nuclear technology. Parties in a position to do so must cooperate in contributing to the development of nuclear energy for peaceful purposes, especially in the territories of non-nuclear weapon states. Under Article I, each nuclear weapon state undertakes, among other things, not to transfer to any recipient whatsoever nuclear weapons or other nuclear explosive devices.

[5] These nuclear facilities and other locations outside facilities included, among other things, power reactors, reprocessing plants, and enrichment plants and were located in 107 countries, plus Taiwan. As a United Nations-affiliated organization, IAEA does not officially recognize Taiwan as a state; however, IAEA applies safeguards there.

[6] IAEA inspectors collect environmental samples from nuclear facilities and other locations, and IAEA's safeguards analytical laboratories analyze these samples to detect traces, if any, of undeclared nuclear material.

[7] LEU is uranium that contains the isotope uranium 235 in a concentration of less than 20 percent and greater than 0.7 percent. Most commercial reactor fuel is enriched to between 3 percent and 5 percent or less of uranium 235.

[8] Throughout this report, we used the exchange rate of $1.3893 = €1.00, as listed in IAEA's annual report for 2011, to convert IAEA budget figures from euros to U.S. dollars.

[9] GAO, *Nuclear Nonproliferation: IAEA Has Strengthened Its Safeguards and Nuclear Security Programs, but Weaknesses Need to Be Addressed*, GAO-06-93 (Washington, D.C.: Oct. 7, 2005).

[10] NNSA is a separate, semi-autonomous agency within the Department of Energy, with responsibility for the United States' nuclear weapons and nonproliferation programs, among other things. NNSA conducts its activities at headquarters and at research and development laboratories, production plants, and other facilities. NNSA also provides technical assistance to IAEA's safeguards and nuclear security programs.

[11] Results from nonprobability samples cannot be used to make inferences about a population, because in a nonprobability sample, some elements of the population being studied have no chance or an unknown chance of being selected as part of the sample. However, the results of such a sample can provide examples of individual countries' views of the three IAEA programs.

[12] Although the information from such a sample cannot be generalized to all such experts, it can provide illustrative examples.

[13] Throughout this report, we use the term "countries" to refer to states recognized by the United Nations. Although nearly all of IAEA's 159 member countries are party to the NPT and are thus subject to its safeguards requirements, IAEA membership is not a prerequisite for a country being a party to the NPT, nor is being a party to the NPT a requirement for membership in the agency. IAEA may implement safeguards in member and non-member countries that are party to the NPT.

[14] Assessed contributions are payments made as part of the obligations that countries undertake as members of the United Nations. The current payment structure for assessed contributions to IAEA is based on the United Nations scale of assessment, adjusted for membership, with a maximum rate (25 percent) and a minimum rate (.001 percent). The scale for IAEA also includes a slight premium to cover the costs of the nuclear safeguards program. For 2012, the U.S. assessment scale rate for IAEA was 25.6 percent.

[15] In addition, a number of member countries make in-kind contributions, such as equipment and expert services, to the agency's various programs on a voluntary basis. For instance, according to State, the United States provided $12.7 million in in-kind assistance to IAEA's safeguards program in 2011, including $2.3 million for environmental sample analysis.

[16] In 1985, the United Nations imposed a policy of "zero real growth" in an effort to stem the upward growth of budgets and to improve the efficiency of the United Nations system across the board. The term "zero real growth" is defined as no growth in the budget beyond that needed to compensate for inflation.

[17] Source material includes natural uranium, depleted uranium, and thorium; special fissionable material includes plutonium-239, uranium-233, and enriched uranium.

[18] GAO-06-93.

[19] Under the small quantities protocol, the reference to small quantities of nuclear material means, among other things, up to 1 kilogram of special fissionable material, 10 metric tons of natural uranium and depleted uranium with an enrichment level above 0.5 percent, 20 metric tons of natural uranium and depleted uranium with an enrichment level of 0.5 percent or below, and 20 metric tons of thorium.

[20] Radiological material—encapsulated in steel or titanium and called a sealed source—is commonly found in equipment used by hospitals and medical facilities to treat cancer patients, among other things. See GAO, *Nuclear Nonproliferation: Additional Actions Needed to Improve Security of Radiological Sources at U.S. Medical Facilities*, GAO-12-925 (Washington, D.C.: Sept. 10, 2012). Radiation detection equipment is installed at border crossings and ports of entry, such as seaports and land border crossings to prevent nuclear smuggling. See GAO, *Combating Nuclear Smuggling: DHS Has Developed Plans for Its Global Nuclear Detection Architecture, but Challenges Remain in Deploying Equipment*, GAO-12-941T (Washington, D.C.: July 26, 2012).

[21] The Nuclear Threat Initiative is a nongovernmental organization focused on strengthening global security by preventing the spread of nuclear, biological, and chemical weapons.

[22] As of April 1, 2013, IAEA reported that the agency had received contributions or pledges to IAEA's nuclear fuel bank from the European Union ($34.7 million), Kuwait ($10 million), the United Arab Emirates ($10 million), and Norway ($5 million).

[23] In addition, Taiwan has implemented the measures of the Model Additional Protocol since October 1999. As noted earlier, IAEA does not officially recognize Taiwan as a state but applies safeguards there.

[24] For more information on U.S. efforts regarding the Additional Protocol, see *Managing Sensitive Information: Actions Needed to Prevent Unintended Public Disclosures of U.S. Nuclear Sites and Activities*, GA0-10-251 (Washington, D.C.: Dec. 15, 2009).

[25] IAEA typically takes several years to reach an initial broader conclusion for a given country and thereafter reevaluates the conclusion on an annual basis. Although the five nuclear weapon states have Additional Protocols in force, IAEA cannot draw broader conclusions for the nuclear weapon states because they have more limited safeguards agreements in force, rather than comprehensive safeguards agreements.

[26] GAO-06-93.

[27] The 13 non-nuclear weapon states that had not brought comprehensive safeguards agreements into force as of the end of December 2012 were: Benin, Cape Verde, Djibouti, Equatorial Guinea, Eritrea, Guinea, Guinea-Bissau, Liberia, Micronesia, Sao Tomé & Principe, Somalia, Timor-Leste, and Vanuatu.

[28] For nuclear weapon states and for countries that are not party to the NPT, IAEA has more limited safeguards agreements. In particular, the agency has limited information concerning nuclear activities in North Korea, which withdrew from the NPT in January 2003. In addition, IAEA has limited information and authority concerning nuclear activities in the five nuclear weapon states (China, France, Russia, the United Kingdom, and the United States) and three countries that are not parties to the NPT (India, Israel, and Pakistan). Specifically, safeguards are implemented in selected nuclear facilities in the five nuclear weapon states, and the three non-NPT party countries have only item-specific safeguards agreements in force and no Additional Protocol in force.

[29] As of early December 2012, the other donors were Canada, the Czech Republic, France, Germany, Greece, Ireland, Israel, Japan, Kazakhstan, Netherlands, Norway, South Korea, Spain, United Arab Emirates, the United Kingdom, and the European Commission.

[30] Among other things, the Department of Energy's Brookhaven National Laboratory manages U.S. technical assistance to IAEA's safeguards program.

[31] Herman Nackaerts, Deputy Director General and Head of the Department of Safeguards, IAEA, "The Future of Safeguards: Adapting to Change," June 7, 2011, and "IAEA Safeguards: Cooperation as the Key to Change," presented at the Institute of Nuclear Materials Management's 52nd annual meeting, July 18, 2011, and published in the *Journal of Nuclear Materials Management*, Fall 2011.

[32] A country is considered to have a limited nuclear fuel cycle if it does not possess all the capabilities of the fuel cycle, particularly the more sensitive capabilities of uranium enrichment and reprocessing, which could potentially be used for nuclear weapons development.

[33] IAEA defines a person-day of inspection as a day during which a single inspector has access to a facility for a total of not more than 8 hours. This definition does not necessarily coincide with a calendar day, so if an inspection requires only a small portion of a calendar day, it still constitutes one person-day of inspection. This metric also does not include an inspector's travel time to or from a facility.

[34] Lawrence Livermore National Laboratory is one of the Department of Energy's national laboratories that conduct various research and development and other activities. Among other things, Lawrence Livermore advises NNSA on matters pertaining to IAEA and is part of IAEA's international network of safeguards analytical laboratories, which analyzes nuclear material samples collected by IAEA inspectors.

[35] GAO-06-93.

[36] GAO-06-93.

[37] In December 2012, in response to a recommendation made by several countries, IAEA changed the name of this database to the Incident and Trafficking Database. To better reflect the nature of the incidents contained in the database, IAEA also redefined the database's goal as "to track incidents of nuclear and other radioactive material not under regulatory control."

[38] In 2002, IAEA established a Nuclear Security Fund to support its nuclear program through extra-budgetary contributions. In 2011, the European Union and 16 member countries, including the United States, donated $25 million to the fund.

[39] GAO-06-93.

[40] In addition, as noted earlier, a number of member countries make in-kind contributions, such as equipment and expert services, to the agency's various programs on a voluntary basis.

[41] GAO-06-93.

[42] GAO-06-93.

[43] GAO-06-93.

[44] Kazakhstan volunteered to host IAEA's fuel bank 20 months prior to the Board of Governors' vote in December 2010 to approve the bank.

[45] The Non-Aligned Movement is a group of 120 countries that consider themselves not aligned formally with or against any international power bloc. The Group of 77 is the largest intergovernmental organization of developing countries in the United Nations. Composed of 131 member countries, it provides the means for these countries to articulate and promote their collective economic interests and enhance their joint negotiating capacity on major international economic issues within the United Nations system.

[46] Article IV of the NPT states, among other things, "Nothing in this Treaty shall be interpreted as affecting the inalienable right of all the Parties to the Treaty to develop research, production and use of nuclear energy for peaceful purposes without discrimination... ."

[47] Of the 35 Board of Governors members at the time of the December 2010 vote to approve the bank, 28 voted in favor of the bank, including 10 members and 3 observers of the Non-Aligned Movement. Six members abstained from voting, all of which were Non-Aligned Movement members or observers, and one Non-Aligned Movement member was absent from voting.

[48] Pacific Northwest National Laboratory, "Options for Creating a Nuclear Fuel Stockpile for Assured Nuclear Fuel Supply," PNNL-SA-48328 (Richland, WA: February 2006).

[49] See, for example, GAO, *Combating Terrorism: Evaluation of Selected Characteristics in National Strategies Related to Terrorism*, GAO-04-408T (Washington, D.C.: Feb. 3, 2004), and GAO, *Rebuilding Iraq: More Comprehensive National Strategy Needed to Help Achieve U.S. Goals*, GAO-06-788 (Washington, D.C.: July 11, 2006).

[50] *Secretarial Determination for the Sale or Transfer of Uranium* (Washington, D.C.: May 15, 2012). Before making any sales or transfers of LEU and natural uranium, DOE is required under the USEC Privatization Act to determine that the transactions will not have an adverse material impact on the domestic uranium industry.

[51] We have previously analyzed transactions that DOE characterized as barters of uranium on two occasions. In both cases, we found that the transactions actually constituted sales of uranium that were authorized under the United States Enrichment Corporation (USEC) Privatization Act but violated the miscellaneous receipts statute. Because we found that the transactions were sales, we did not consider and did not decide whether barters are also authorized under the USEC Privatization Act. We also did not perform such an analysis in this case. See GAO, *Excess Uranium Inventories: Clarifying DOE's Disposition Options Could Help Avoid Further Legal Violations*, GAO-11-846 (Washington, D.C.: Sept. 26, 2011); and GAO, *Department of Energy: December 2004 Agreement with the United States Enrichment Corporation*, B-307137 (Washington, D.C.: July 12, 2006).

[52] In December 2008, DOE published the *Excess Uranium Inventory Management Plan* "to provide the general public and interested stakeholders more specific information and enhanced transparency with respect to DOE's preliminary plans for its excess uranium transactions." The plan identified excess uranium inventories and specific transactions that were planned or under consideration at that time, or that might be considered by DOE in the future, for disposition. In early March 2013, NNSA officials told us that DOE's Office of Nuclear Energy was preparing an updated version of this plan.

[53] GAO, *Nuclear Weapons: Actions Needed to Identify Total Costs of Weapons Complex Infrastructure and Research and Production Capabilities*, GAO-10-582 (Washington, D.C.: Jun. 21, 2010).

[54] In this report, and as we have done in previous reports, we use the term "overlap" to refer to circumstances when multiple agencies and programs have similar goals, engage in similar activities or strategies to achieve them, or target similar beneficiaries. The presence of overlap can suggest the need to look closer at the potential for unnecessary duplication. However, our review did not go to sufficient depth to determine whether and to what extent the programs are actually duplicative, which occurs when two or more agencies or programs are engaged in identical activities or provide the same services to the same beneficiaries. For more information on overlap and duplication in federal programs, see GAO, *Opportunities to Reduce Potential Duplication in Government Programs, Save Tax Dollars, and Enhance Revenue*, GAO-11-318SP (Washington, D.C.: Mar. 1, 2011); and GAO, *Managing for Results: Using the Results Act to Address Mission Fragmentation and Program Overlap*, GAO/AIMD-97-146 (Washington, D.C.: Aug. 29, 1997).

[55] See GAO, *2012 Annual Report: Opportunities to Reduce Duplication, Overlap and Fragmentation, Achieve Savings, and Enhance Revenue*, GAO-12-342SP (Washington, D.C.: Feb. 28, 2012) and GAO, *Nuclear Nonproliferation: Action Needed to Address NNSA's Program Management and Coordination Challenges*, GAO-12-71 (Washington, D.C.: Dec. 14, 2011).

[56] Results from nonprobability samples cannot be used to make inferences about a population because in a nonprobability sample some elements of the population being studied have no chance or an unknown chance of being selected as part of the sample. However, in this case, the results of such a sample can provide examples of individual countries' views of the three IAEA programs.

[57] GAO-06-93.

INDEX

C

D

H

I

J

K

L

M

P

Q

R

S

T

U

V

W

Y

Z